Growing Healthy

How to make over $100,000 p.a. from 5 acres with organic fruit and vegetables (or $30,000 p.a. on 1 acre).

By
Geoff Buckley.

Growing Healthy

Copyright 2009 by Geoff Buckley

Published by Castelen Press
3 Golfcourse Road.
Mt.Tamborine,
Queensland 4272
Australia

Website: www.growinghealthyorganicfood.com

Third Edition Published May, 2010.

ISBN 978-0-646-50320-2

Foreword

Geoff Buckley's book "Growing Healthy" provides impetus to a grassroots movement that promises to change the way we eat and to redefine the way we view food production. I believe that we are entering a golden era in agriculture where food growers will finally be recognised as the most important professionals on the planet. With this recognition will come a sense of responsibility as food producers realise that the health, happiness and longevity of consumers is directly related to their produce. We are what we eat and what we eat comes from the soil. A soil that produces nutrient dense, medicinal food is a living soil, teeming with billions of micro-organisms and rich in the full spectrum of minerals that are the building blocks for protective phyto-chemicals. In this book Geoff reveals the secrets of the sustainable production of high quality food with forgotten flavors and extended shelf-life. "Growing Healthy" contains a host of practical solutions garnered from the coalface. Geoff and Bev Buckley are the founders of a multi- award winning food farm on Tamborine Mountain in South East Queensland. I had the pleasure of training this passionate pair when they attended my four day Certificate In Sustainable Agriculture course several years ago. They arrived at the course with no pre-conceptions and, as such, they were perfectly positioned to embrace these exciting new concepts and make them their own. Since that time they have become educators and innovators in their own right. In fact the Buckley's have become key strategists for the production and marketing of chemical-free, superior food. I trust that you will find something of lasting value in these pages and perhaps you will feel motivated to join the crusade for nutritious, chemical free, locally grown food.

Graeme Sait

CEO, Co-Founder
Nutri-Tech Solutions Pty Ltd
PO Box 338
Eumundi Qld 4562
Australia
Ph: +61-7-54729900

Introduction

Food shortages have started and as the world population increases this will only get worse. The world's population in June 2007 was approximately 6.6 billion, and this is projected to grow to 8 billion by 2030 and 9 billion by 2050. More people need more food and many countries already depend on food aid or imports. With the oil price escalating, the cost of transporting food is becoming more and more expensive, and more countries will not be able to afford it. The logical solution is to grow more food locally and to become more self sufficient.

Several other trends make the situation even more critical. The trend towards cities and away from the country continues. There is spiralling growth of megacities. This means less people on the land to grow food and as cities expand there is less land to grow food on. Just here on Mt.Tamborine we can see houses going up all around us with the rapidly growing population in south-east Queensland. Tamborine Mountain used to be totally rural. It's not really so long ago the population was less than 1000, now it is over 6000. You can count the number of large farms left on your fingers.

The world has an epidemic of diseases such as diabetes, cancer, obesity, heart problems, Alzheimer's and Parkinson, ADD and so on. These are caused by the pollution of our air, water, soils and seas and then resulting poor nutrition. The solution is not in man-made drugs, it is in **growing healthy** nutrient-rich food.

We have climate change which makes it even more difficult for farmers to cope. Just in the past twelve months we have had one of the coldest winters (in 2007) with severe frosts which caused even avocados to drop to the ground in one orchard and this was followed by what was one of the wettest summers with little sunshine for three months. This caused a drop in production because lack of sunshine means lack of photosynthesis which means plants lack energy. The persistent heavy rain also meant that we lost 2 or 3 months of planting time.

The good news is that organic farming according to nature can overcome these problems because a healthy soil and healthy plants have a strong immune system and can survive and flourish. One of our most experienced local farmers says that he often has his best results when there are bad years such as 2007. This is partly because problems often cause price rises but primarily because he continues to get higher production even in bad years. In 2007 he had an increase of 32% in the quantity of avocados produced on his property compared to 2006.

Mankind has created these problems and therefore mankind can also solve them. There is an increasing awareness that we have to reverse some of these trends if we are not to destroy our planet. There are increasingly individuals and groups speaking out and acting to help change the world into a better place. One of the early ones was Fritz Schumacher with his "Small is Beautiful" published in 1974. He had many wise sayings including "People who live in highly self sufficient communities are less likely to get involved in large-scale violence than people whose existence depends on world-wide systems of trade." Another leader is Judy Wicks (see www.judywicks.com & www.whitedog.com) from Philadelphia, USA. She has started a world-wide movement based on the premise that an environmentally, socially and financially sustainable global economy is comprised of **a network of sustainable local economies** – or living economies that support natural life and community life. Rather than a global economy controlled by large corporations, she envisions a decentralized global network of local living economies, comprised of independent, locally owned businesses - creating community vitality and wealth. Basic needs are produced locally and what is not available is bought from other local communities where the products originate (see also www.livingeconomies.org). I strongly recommend you join or start a local farmers' market. People go to farmers' markets not just to get food, but for the sense of community that comes from it. Ours is a real social event. As Judy Wicks says **"happiness comes from belonging, not belongings".**

As transport is critical to our current food system, as oil prices rise this is a compelling additional argument for localization rather than globalization.

We need to think about changing our business model so that instead of basing it on growth and competition, we focus on increasing **quality not quantity and cooperation not competition.** Business is not all bad, it can be good and beautiful and satisfying, but one of its major goals needs to be not maximising profit but service to its customers and to the community. Serving the interests of its shareholders is not the right primary focus: customers and community are at least equal if not more important. This is easier to achieve in small local businesses than giant corporations—small really is beautiful. Instead of an international stockmarket it would be a better philosophy for people to invest in their local businesses to help people they know and that makes life better where they can see and have a say in the results themselves.

Much of the world's inequity stems from the power wielded by the large (often multinational) companies that control governments; it would be great to be able to limit the size of companies but a more effective practical solution is to support smaller local companies that help the local communities. If we **buy local**, we will eventually have an impact on the multinationals.

Business needs to move away from the old paradigm of financial growth to measure success; it needs to grow deeper rather than bigger, to increase quality, to increase relationships and to invest in local communities. Capitalism is focussed on numbers and needs to wake up to realising that success depends on the longterm impact on people and the planet, not just on short term profit.

Geoff Buckley, November 2008.

"Growing Healthy"

How to make over $100,000 p.a. from growing fruit & vegetables organically on 5 acres.

Contents

Foreword
Introduction
Acknowledgments
About this book
List of 48 photographs

1. Why grow your own food?
    *   Ten compelling reasons
2. What farming method will you choose?
3. Principles of Success
    *   Fourteen Key Principles
4. How I became a farmer.
5. Choosing a property
    *   Eight factors to consider
6. Farm layout
    *   weatherproofing your farm
    *   protection against frosts
    *   the use of greenhouse or shadehouses
7. Getting started.
8. What are the principles to use to decide what to grow?
    *   four key criteria
9. Deciding which vegetables to grow
    *   crop rotation
    *   seeds v. seedlings
10. Deciding which fruit to grow
11. Deciding which herbs and spices to grow
12. Deciding which nuts and berries to grow
13. Soils, roots and leaves
    *   your soil is your major asset
    *   what are soils made of?
14. How to grow crops
    *   the five things you need to grow anything
    *   nature's two magical processes that do most of the work
    *   the magic formula to avoid pests and diseases
15. Soil building activities: mulching / mounding / composting
    *   the importance of mulch

Acknowledgments.

I would like to acknowledge two people without whose input this book would not have been possible.

Firstly my wife, Bev, my partner in marriage and in the business. She found our property on Mt.Tamborine and persuaded me to buy it. It took her two years to do so! But it was one of our best decisions ever. She also thought up the idea of starting the Tamborine Mountain weekly farmers' market, which has been a huge success both for the community and us. Bev has supported me in growing the business and in making it happen. Bev also writes our monthly newsletter from which I have borrowed freely (see www.greenshed.com.au or http://healthyorganicnow.com for our newsletters).I am sometimes amazed at her energy and the work she has done. I am forever grateful, because I love what we do and what we have achieved together.

Secondly, I thank Graeme Sait, CEO of Nutri-Tech Solutions Pty.Ltd. of Yandina on the Sunshine Coast. It was his talks to our Tamborine Mountain Local Producers Association that led us to do his "Certificate of Sustainable Agriculture" course which taught us most of what we know about farming. But most of all he kindled in me a passion for agriculture and farming. I am proud to call myself a farmer.

I would also like to acknowledge three other experts from whom I have learned much:
1.Ken Pegg, from the Queensland Department of Primary Industries, who is one of Australia's authorities on avocados and who has patiently and clearly answered many of my questions about avocados. 2. Peter Andrews, the author of "Back from the brink", the most important book on Australia's soils that I have read. His research is a major source of much of the material that makes up chapters 3 and 15. I recommend his book to all who really want to understand the soil and water. 3. Gary Zimmer, the author of "The Biological Farmer," who kindly gave me permission to include a quote and diagram on the nitrogen cycle. I have read his excellent book, listened to him talk and watched him on videos. He has a passion for farming and his book is an excellent practical guide of value to all farmers.

About this book

There were **two** aims in writing this book. **One** was to produce a textbook for those who do our training course which we call "Growing Healthy"—teaching people how to grow their own fruit and vegetables organically. My wife, Bev, and I have both had many years' experience in education and training and we have learned that the best way to learn is by doing. We provide practical work in our course so that people can try things out and see the results of what they do. The other big lesson for us has been that rather than try to persuade someone about something new, the best way is to demonstrate. So we have created a model farm on five acres to demonstrate how much can be produced in a small area. One idea we teach is hard for people who have any experience with gardening or farming to accept. This is what we now call our magic formula to deal with insect pests, animal and bird pests and diseases. We do not use any pesticides or herbicides; we use our magic formula which is explained in the book and which we are happy to demonstrate.

The **second** objective is to demonstrate that you could make a good living on five acres. Hence the subtitle for the book—to demonstrate that you can make over $100,000 p.a. on five acres. With a smaller area you can make say $30,000 p.a. from one acre. Writing the book made me realise I had gone along without a clear business plan and it also caused me to rethink the factors that make for profitability.

We have in practice taken much longer than necessary to achieve this objective. You can learn from our experience and I provide the steps for you to take to achieve it more quickly than we have done.

Major goals are not all financial. We want to create a beautiful property where we can experience nature and be at peace. We want to be as self-sufficient as possible and we grow some things just for their beauty, fragrance and health-giving properties such as herbs, flowers and some trees.

One of the most important messages in this book is the significance of the major elements in the health of the soil, plants and humans. There are 92 elements that make up the earth and the human body. Not a lot is known about many of these elements but I believe they all have an important role. However there are eighteen that crop up time and time again throughout this book. To save continually repeating information or referring back and forth, I have put a summary about each of the eighteen into Appendix A. This is a key reference.

To achieve the financial goal in the book's subtitle, the most important decision is "what crops do I grow?" I did not realise the significance of making good choices for many years so my experience should be a valuable lesson for you.

44 Photographs

1.Bev & Geoff winning small business champion award for fresh food

Our Property

2. Our vegetable garden looking north.
3. Plum tree
4. Dam on our stream
5. View from our balcony
6. View from training centre building site

Planting

7. Seedlings
8. Strawberry plants in pots
9. Strawberry plants in garden
10. Strawberries growing

Rhubarb

11. Rhubarb growing
12. Rhubarb in transit to packing shed (Bev on rideon)
13. Rhubarb picked
14. Rhubarb packed

Lime trees

15. Lime trees: 100 ready to plant
16. Preparing soil for limes—mulch and cardboard
17. Limes just planted
18. Limes showing tall sprinklers
19. Twelve year old lime.

Other fruit trees

20. Tamarillos
21. Macadamias—3 years old
22. Macadamia—12 years old
23. Citrus-Lemons

Avocados

24. Avocados in the mist
25. Avocado in flower with small sprinkler

Examples of different crops

26. Turnips
27. Kale
28. Woofers picking beans

Blueberries

29. Blueberries recently planted
30. Blueberries for sale at market

Equipment:

31. Tractor with slasher & woofer
32. Trailer with canopy behind our stationwagon
33. Fertigation unit

34. Foliar spray unit

Market Day

36. Getting ready for the market
36. Produce in trailer ready for market
37. Woofers picking avocados for market
38. Carrots and silver beet at our market
39. At the market
40. Flowers at market
41. Woofers washing ginger

Sheep

42. Our ram with his harem
43. Rideon with trailer & sheep
44. Our first lamb.

Chapter 1
Why grow your own food?

"Agriculture…is our wisest pursuit, because it will in the end contribute most to real wealth, good morals and happiness." Thomas Jefferson to George Washington, 1787.

Earth and the human body are made up from 92 elements. These are however not spread evenly around the world. Gold does not occur everywhere. Selenium which has been shown to combat cancer and AIDS is almost non-existent in Australia's soils. Water, which consists of hydrogen and oxygen, is on 70% of the earth's surface and the body is 70% water, but water is relatively scarce in Australia when there are floods in USA and parts of China right now. The elements needed to grow nutritious fruit and vegetables are not always present in the soil. **Plants cannot create them**. The result is that while broccoli is considered generally a highly nutritious vegetable, in practice its nutrient content depends on the particular soil it was grown on. If an element is not in the soil it won't be in the fruit or vegetable. Either this fact is not commonly understood or people are choosing to ignore it.

Conventional or "industrialised" farming methods produce over 90% of our food using large quantities of pesticides, herbicides, insecticides and man-made chemical fertilisers that are both toxic and generally contain only a few of the natural elements. Since World War Two the so-called "green revolution" has advocated N-P-K fertilisers, which as the name suggests primarily contain nitrogen, phosphorus and potassium. Calcium, magnesium and all the multitude of trace elements which have been proven to be necessary for a healthy balanced soil and nutritious food have been ignored. For this reason the quality of the world's farmlands has been declining steadily since the 1950s and organic matter has been destroyed by all the ever increasing quantities of toxic chemicals used in conventional farming. Organic matter 50 to 60 years ago used to average 5% of our soils. Today they average 1 to 1.5 %. What was once live soil full of billions of microbes, bacteria, worms and fungi is becoming dead soil. Industrial farming methods lead to soil erosion, salinization, desertification and loss of soil fertility. Chemical farming does however produce large **quantities** of food. Unfortunately quality has been forgotten.

There are ten compelling reasons to grow your own food.

1. Growing your own food provides yourself and your family with nutrient-rich health-giving food to replace the food available generally, which lacks nutrients but does have plenty of toxic chemicals sprayed on it regularly from its inception through till picking. Do you know that tomatoes are our most popular food and they are typically sprayed with six chemicals every three days of their life? Oranges do not always contain vitamin C.

Another reason that our food is being ruined is in its processing, storage, transport and preparation. All four of these steps cause nutrients to be depleted. Nutrient content generally declines over time, so buying food transported round the world and imported defeats the purpose if you are looking for food as a source of nutrition. Why buy food transported thousands of miles from WA, Tasmania or Victoria to Queensland? It makes sense to buy food grown locally and best of all in your own garden where you know what goes into it. Remember good food tastes better.

2. The cost of food is another factor. Uncertain weather patterns or climate change are causing dislocation of food supplies. Flooding is destroying crops. Unseasonal hot periods are damaging crops. Droughts are reducing yields. With rising oil prices our food is going to increase in price, because the cost of inputs is going up.   It is not only transport of food, but the growing of food that is dependent on oil.

    Large industrialised farms use oil and diesel for their machinery, and the man-made fertilisers, herbicides & pesticides used in ever increasing quantities are largely petro- chemically based. With oil in short supply and rising in price, our farmers either will go broke or have to put their prices up or receive more subsidies.

    The growth of mega-cities also contributes to food shortages. The whole world is becoming increasingly urbanised and as cities grow in size, this decreases the land and people available for farming. Australia today is the most urbanized country in the world. Its five major cities account for 58% of the population. The largest 11 Australian cities account for 71% of the total population (from an analysis of ABS statistics).

    In recent years competition for food crops has increased because of the demand for crops such as corn and soy beans which are used in the production of ethanol and bio-diesel. Corn and soy bean prices have doubled and trebled in the past two years.

3. Tinned and processed foods generally have preservatives in them to extend the shelf-life and kill microbes and bacteria. Many processes remove enzymes and nutrients. Preservatives' job continues in the human body when you eat these foods i.e. the preservatives kill microbes and bacteria in your body. These "bad guys" that are being killed are really "good guys" in terms of digestion. Your intestines are supposed to have billions of them to properly digest the food you eat so that you can absorb any nutrients in the food. When we kill them, we get sick. We get stomach pains, indigestion, cancer, inflammation, flatulence and constipation because there are no bacteria and microbes left to carry out the digestive process. Today one of the biggest growth areas in the medical industry is looking after digestion problems.

4. With bought foods you get hundreds of different additives, flavourings and colourings. There are over 74,000 man-made chemicals used in commerce with about 2000 new ones being added each year. There are only 92 elements that make up the earth and the human body. So how is our body going to recognize over 74,000 new ones and how can we expect it to deal with them? Whether we are aware of it or not, we are being constantly bombarded by a host of chemical toxins which are affecting our health. The most insidious of these are the ones sold to us as "food." Packaged foods have a series of codes on the labels that list the "non-food" content in what we are buying. The value of the book "The Chemical Maze" (see reference 1) is that it enables you to find out what the numbers refer to and which are harmful. It lists 296 additives (including the 25 that are allowed in Australian foods but are banned in countries overseas), rates them and details health problems that they may cause. One would expect that a cereal bar, sold as a school lunch box food by a well known cereal manufacturer, would be beyond reproach. The reality is somewhat different from the image of wholesomeness, health and vitality which this company promotes in its advertising. This particular cereal bar contains seven chemicals described as numbers 492, 322, 171,102, 133, 124 and 472e. There is no indication as to what these numbers represent which makes the information worse than useless. Both 124 and 133 originated as coal tar dyes. Additive no. 102 is "Tartrazine" and is thought to cause problems associated with asthma, headaches, concentration problems, depression, learning difficulties, behavioural problems, insomnia, confusion and on and on. These additives are banned in countries overseas and are thought to be carcinogenic. It is not surprising that the company that manufactures this "food" lists the additives it uses in a type size that is so small that it takes a magnifying glass to read it.

You are responsible for your health. The medical profession is generally too busy looking after sickness, treating symptoms with drugs and surgery. The use of antibiotics is prevalent and as the name suggests these kill off the microbes and bacteria in your body, the necessary beneficial ones get killed plus the few harmful ones. Do the doctors tell you what you need to do to put them back?

Dr. Linus Pauling, the only person to win two Nobel Prizes on his own, said all illnesses and diseases can be traced to mineral deficiencies. Dr. Willem Serfontein said: "Almost all diseases and afflictions, both mental and physical, can be traced back to nutritional deficiencies. The answer to optimal physical and mental health therefore lies in harnessing the power of nutrition." (reference 2).

The world has an epidemic of health problems with increasing rates of cancer, obesity, diabetes and heart problems. Nutrient-rich food without herbicides and pesticides is the solution, but nearly all the world's research is focussed on finding patentable drugs to sell to help alleviate the symptoms. There is finally a reluctant acceptance that if you eat organic food you will live longer and in better health. A major study funded by the European Union concluded that organic food is more nutritious than ordinary produce and that it may help to lengthen people's lives.

5. Fresh food makes business sense. In Australia it is GST free and generally a cash business, so paperwork is much less than most businesses e.g. no invoices and no debts to chase. Any food that's not sold can always be eaten, swapped with other suppliers, composted or made into jams or chutneys. Food prices are likely to continue to rise, because of oil price rises and transport costs. Newly affluent Chinese and Indians have acquired a taste for Western style food. Bread, beef, pork and noodles etc. are increasingly in demand. This is putting yet another increased pressure on supplies.

Farmers have repeatedly been told "to get big or get out" but the facts tell us a different story. Which is more profitable, a small family farm or big industrial farms? The surprising fact is that according to USDA (USA Dept. of Agriculture) records from the 1990s, small family owned and run farms have a higher average net profit per acre than industrial farms. Per-acre-profit declines steadily as farm size grows. This is because small farms:
1. use each square metre of land more intensively and
2. grow a more diverse selection of produce suitable to local food preferences and
3. sell more directly to the consumers the average price is closer to retail than wholesale. (Reference 3 "Animal, Vegetable, Miracle" by Barbara Kingsolver page 76. See also www.nffc.net).
We have been deluded into thinking that big is better and more profitable. Small can be beautiful.

6. Growth in demand for organic produce exceeds supply and has been growing at 20% per annum. Price premiums are common as the labour component is greater. The labour element is to my mind a plus in the broad scheme of things because industrial farming and industrialization generally uses technology to reduce employment, so organic farming helps to provide employment, and of course helps the environment too by reducing CO_2 emissions.

7. A decrease in the world bee population lowers pollination levels, which means lower productivity. This is another factor why world food production is under threat-- so grow your own. The internal alarm of the entire species of bees has gone off and they are responding by abandoning hives. Bees are removing themselves from poisoned agricultural areas and are moving; sometimes great distances to find healthier environments to try ensure the future of their species. The bees' message is that the continuous manipulation of natural systems cannot continue.
Isaac Newton said if we lose bees, mankind will last only a few months!

8. Farming takes you back to nature and away from the concrete jungle. All of us need to feel the fresh air and open spaces and find a quiet spot with beauty, peace and stillness. That is why most people love trips to the country, picnics, golf, farm stays and so on, as city life gets more and more stressful and unhealthy.
A study on why people chose to farm found their main reasons were:

(i) I like to work outdoors;
(ii) It's a good place to raise my children;
(iii) I'll always have a place to live and food to eat.

9. Growing your own food is a very rewarding activity. It's the first time in my life I can say with total conviction "I love what I do." Our weekly farmers' market creates a real sense of community and belonging. It is a social event.
    Happiness comes from belonging not belongings. It can also be financially rewarding as this book demonstrates.

10. We all strive to find a worthwhile purpose for our lives. Food is one of the basic necessities of life. Farming is without doubt a worthwhile activity. What better purpose than to leave a well cared for piece of land for your children and posterity?
    The Indian philosopher, Saint Basava, said there can be nothing more sacred than work. Thomas Aquinas said: "There can be no joy in life without the joy of work." I can assure you farming is work! You want to sleep soundly? This is the solution.

Chapter 2
Different farming methods to choose from.

"Try not to become a man of success, but rather try to become a man of value".—Albert Einstein.

A major consideration when deciding how to grow crops is what approach to farming to take. There are at least five different methods and these will each take you down different paths:
1. Conventional farming
2. Permaculture
3. Biodynamics
4. Organic farming, with or without certification
5. Biological farming or Nutrition Farming.

Conventional farming

Farming up till about 1950 was fairly soundly based on the experience of hundreds of years. Since then the impact of large multinational petro-chemical companies has changed the face of farming throughout much of the world with the so-called "green revolution". After over 50 years of this new approach with man-made chemicals, fertilisers, seeds, pesticides, herbicides, genetic engineering and so on, farmers are starting to become aware of the enormous longterm damage that has been done to our soils, the quality of our food and to human health. There are over 74,000 man-made chemicals now floating around our world helping not only to destroy our soils but also polluting our air and water, the sea and the earth's whole environment. Pollution is everywhere and the earth and its people are rebelling. Our bodies cannot cope with all these chemicals—they and the earth are made up of relatively few natural elements and they do not know how to cope with this invasion of new substances. The media wonder why there is an epidemic of new diseases and obesity. The multinationals sell ever increasing quantities of pesticides and herbicides but the pests keep coming. The multinationals sell ever increasing numbers of drugs to humans to combat the health problems they create.

One of the major solutions to many of the world's problems is to revert to farming that does not depend on pesticides and herbicides and man-made chemical fertilisers. Instead of focussing on increasing quantity of production we might focus on the quality of our food and the longterm sustainability of our soils. Conventional farming reduces the organic matter in the soil (by killing off the soils' life with herbicides and pesticides) and this releases CO_2 into the atmosphere. Various people have estimated that farming contributes between 35 and 55% of the CO_2 that is a major contributor to climate change and temperature increases (reference 1). Cattle contribute methane gas which is more harmful than CO_2. Organic farming reverses this trend by increasing the levels of organic matter in the soil and reducing CO_2 in the atmosphere and could be the fastest way to reduce the impact of humans on climate change. Governments ignore this fact e.g. in the latest carbon reduction plans agriculture is exempted, I wonder who influences their decisions? You might guess that I am not in favour of conventional farming.

Permaculture

Permaculture evolved in the 1970s with work done by Bill Mollison and David Holmgren. It started with the belief that people need to feed themselves sustainably and need to move away from reliance on industrialised agriculture using fossil fuel focussing on high yields of a single crop (on quantity not quality). The model is based on diversity and an abundance of small scale market and home gardens for food production with food miles being a primary issue. "Permaculture One" was published in 1978.

Howard T. Odum espoused the maximum power principle which examines the energy of a system and how natural systems tend to maximise the energy embodied in a system e.g. the total calorific value of woodlands is very high with its multitude of plants and animals. It is an efficient converter of sunlight into biomass. A wheat field on the other hand has much less total energy and often requires a large energy input in terms of fertiliser. Esther Deans who pioneered the No-Dig Gardening methods was another influence.

Modern permaculture is like a system design tool. It is a way of
- looking at the whole system
- seeing communications between the elements or parts
- observing how the parts relate
- planning to mend sick systems by applying ideas learned from longterm sustainable working systems

We investigated permaculture early on in our learning period and decided it was not for us—too idealistic and not practical enough. The key principles are sound enough, but we wanted a more specific "how to" approach.

Biodynamics

This is another approach which has taken off in many areas and become a strong movement with its own organisation and good results. There are quite a few courses in biodynamic farming and it is sometimes called an advanced form of organic farming.

It is based on work done by Rudolf Steiner who is responsible for many novel ideas, including the Steiner Method of education for our children. He was Austrian and lived from 1861 to 1925.

His first principle was that a farm should be self-sufficient and bring no (or few) outside materials onto the farm but produce all needed materials such as manure and animal feed from within. He adopted planting according to the cycles of the moon and the plants. Making compost in a specific manner is also an important factor. There is also a special use of cow manure.

There is an impressive CD/DVD (called "How to save the world"—available from www.biodynamics.net.au) about the impact Biodynamics is having on India, helping to reverse the 50 years of negative impact on farms and farmers of the multinationals. Apparently the suicide rate among Indian farmers who saw their farms becoming less and less productive and presumably smaller and smaller was huge. Biodynamics has helped provide some hope and has caught on because it advocates the use of cow manure and cows are sacred animals in India.

We stayed with a farmer in Margaret River, W.A. recently and he has changed to Biodynamics after many years of conventional farming and it was great to see his enthusiasm and excitement at seeing improvements in his farm and the quality of food produced. Many farmers have experienced the negative health impact of using toxic chemicals and are thrilled to work with nature once more.

We did a workshop on biodynamics very early in our search and decided it was not for us. The burying of cow horns and planting according to the cycles of the moon seemed to be too "New Age" for us at the time. Now that we know more about farming we think biodynamics is a lot sounder than we thought earlier. The use of natural manures and composting is clearly excellent and our friend in Margaret River uses fertiliser products from the sea to ensure all the trace elements are present, in addition to the other Biodynamic formulas such as compost preparations and horn manures.

Organic Farming

The term organic is widely misunderstood and its use has come to mean many things. The primary meaning of organic farming is that it means farming without the use most pesticides and herbicides and this is a good thing. Who wants toxic chemicals sprayed on their food? However there are five or six different organisations "certifying" organic farms and charging large annual fees to audit farming practices used. It can take three years to obtain certification. It seemed to us that they have a fairly negative approach, telling you what **not** to do. We thought we knew what not to do and were looking for an approach that told us what **to do**.

Certified organic food can be free of toxic herbicides and pesticides but have no nutrients in it. Because the essential point is to grow food that is nutritious which means you need to ensure that you have all the elements in your soil. Plants cannot create elements that are not in their soil, so broccoli is supposed to have lots of iron and other nutrients but if it is grown in soil that is deficient in them it will lack them.

Most food produced without herbicides and pesticides sells really well with or without certification. Certification seems to me to be an excuse to charge higher prices; certified produce generally sells for between 20 and 100% more than conventionally produced food. As organic farmers do not use increasingly expensive petro- chemically based fertilisers, pesticides and herbicides I wondered whether the price should be lower!

The main argument for higher prices is the greater use of labour to remove weeds and the fees to be certified. There are techniques to control weeds, such as the heavy use of mulch.

Anyway we decided that we would not seek certification. I grow crops the way Nature approves. I do not like the idea of having to pay someone to approve how I do things when it is already approved by nature as is evidenced by the health and vigour of the plants and trees and the absence of diseases and pests. In practice not having certification has not been a problem – the demand for fresh nutrient-rich great tasting produce that has no toxic sprays far exceeds our capacity to produce.

If you are interested in obtaining organic certification, the cost of being ready for certification will depend on the prior uses of the property you choose. If the land has been used for conventional farming, you will need to take measures to eliminate the toxicity and rebuild the organic matter and soil life. This all takes time and money. If in doubt, you should get the soil tested for toxicity.

Biological farming or Nutrition farming*

The term Nutrition Farming is a registered trademark of Nutri-Tech Solutions P/L (NTS): "This approach emphasises the production of nutrient-rich food with forgotten flavours and greatly extended shelf-life. Nutrition farming is working with nature rather than against, where productivity, profitability and sustainability are seen as inseparable." It assumes no use of pesticides and herbicides and moves on from there to explain why and how to grow healthy nutrient–rich plants.

We were fortunate to come across Graeme Sait, CEO of NTS who are based at Yandina on the Sunshine Coast. He teaches this approach around the world and we did his "Certificate of Sustainable Agriculture" course in 2004. Nutrition farming is similar to the more general biological farming approach focussing on regular testing and balancing the soil, and feeding and encouraging the multiplicity of soil life (bacteria, worms and fungi).

It is based on the seminal work done by Professor William Albrecht, who died at the age of 85 in 1974. He studied soils in the USA, UK, Europe, Egypt and Australia. He published many papers and his work has been put together in four volumes called "The Albrecht Papers". He has many famous sayings. One of my favourites is "Insects and diseases are the symptoms of a failing crop, not the cause of it." Albrecht's work has been carried on by others such as Dr. Carey Reams, Neal Kinsey, Dr. Arden Andersen and Graeme Sait, all of whom are authors of great books on soils and agriculture.

Graeme runs a four or five day intensive certificate course both in Australia and overseas. I highly recommend you do this course if you are serious about growing your own food profitably and enjoyably. We also run courses that emphasise the basics and provide practical demonstrations of how to get started and what to do. We get participants to have a go, as we believe in learning by doing. See Appendix G for details of training courses.

We are very comfortable with the nutrition farming approach and after doing his course I could see the difference in our results within months and became confident that we can grow anything! We try to demonstrate this by growing 60 different crops.

Chapter 3
Principles of Success

"To laugh often and much; to win the respect of intelligent people and the affection of children; to earn the appreciation of honest critics and endure the betrayal of false friends; to appreciate beauty; to find the best in others; to leave the world a bit better, whether by a healthy child, a garden patch or a redeemed social condition; to know even one life has breathed easier because you lived. This is to have succeeded."—Ralph Waldo Emerson.

The three pillars of sustainable agriculture are:
- Biodiversity
- Mulching
- Do not plough or clear the land.

In addition to these three principles I have distilled the major lessons we have learned over the past ten years into this chapter to provide a foundation for discussing how you can be highly successful in growing your own fruit and vegetables organically.

1. Biodiversity.

By having a large range of crops you minimise the risks to your business. It reduces the risk arising from price fluctuations. It reduces the impact of climate change and also the risk from attack by pests and diseases. If you follow the conventional farming path of specialising in 1 to 4 crops or animals, you have a bigger area to attract pests, viruses and diseases. We have seen farmers lose most of their income in a year because the crop is wiped out by a disease or virus.

Biodiversity also adds to your interest and excitement. Humans need variety in their food and clearly you will benefit financially by lowering your own food budget as you grow more of what you eat. Last but not least your health will benefit too.

2. Mulching.

Soil preparation before planting is one of the keys to success. Mulching is part of this process but is also vital on an ongoing basis. The reasons why mulching are so important are covered in detail later. We go through large quantities of mulch in our vegetable area and also when we are preparing soil for planting new fruit trees or to help in keeping weeds down around existing trees.
- Sources of good mulch are weeds. Weeds drag nutrients up from below and when they die or are removed the nutrients should be put back as mulch or compost into the topsoil to improve soil fertility.

- The same should happen to grass cuttings and the remnants of fruit and vegetables, such as the roots and stalks of broccoli and cauliflowers when you have picked the crop.
- Pruned leaves and branches are good sources of mulch. We also get loads of mulch from the local tip's green waste area, where a contractor mulches up into woodchips all the branches and leaves taken to the tip. Woodchips take longer to breakdown than most other mulches but are good at keeping weeds down.
- Hay bales are excellent but hard to get in our area
- We use a lot of sugar cane mulch which we buy for $3.50 a bale.

3. No ploughing or tilling.

We do not have or use a rotary hoe or plough. Land clearing needs to be kept to a minimum. What we do when we plant in a new area that is covered with grass &/or weeds is a three step soil preparation process:

- Spread fertiliser directly onto the grass. We usually use three or four items such as chicken manure, rock dust, gypsum and a general balanced fertiliser with trace elements such as Nutristore Gold (from NTS).
- Put down 2 or 3 layers of cardboard &/or newspaper to block out the light. These will break down fairly quickly. We collect cardboard from the local hotel.
- Put a thick layer of mulch on top of the cardboard. This is not only good for all the reasons mentioned above but stops the cardboard being blown away. (See photos 18 & 19).

This process takes time to do and time to break down all the materials. Ideally you need to prepare ground like this months or even a year before planting. The cardboard and mulch and dead grass help create organic matter promoting soil life and improving the water holding capacity.

4. Strive to continually improve your soil fertility.

You have to continually work at maintaining and improving soil fertility. Once you achieve a balanced soil and a pH around 6.4, you have to realise that the plants you grow will remove nutrients and rain and irrigation will leach out some of the elements. Boron, magnesium and sulphur are readily leachable and need replacing at least once every year.

If you test your soil pH in several places you are likely to find variations not only on a macro scale but also on a micro basis. There will be pockets that are poorer than the rest. So it is an ongoing task to focus on improving your soil. This means more than just adding the natural elements, it means maintaining and improving the level of organic matter, the microbes and fungi.

You might find that drainage needs to be improved.

Allowing weeds to grow can also add to soil fertility, because they drag nutrients from low in the soil and tree leaves and weeds when they die after a relatively short lifecycle provide a nutrient-rich mulch.

Weeds are nature's way of putting soil fertility back. The fact that they exist shows the soil lacks fertility. They add organic bulk back to the soil, grow rapidly and some deter animals from grazing, thus enabling the soil and other vegetation to recover. For example, thistles produce an extremely large green surface area and so accumulate fertility faster. Thistle seeds have a higher concentration of minerals and nutrients than almost anything else. Grass takes five to six months to grow and complete its life cycle. Weeds grow in as few as 40 days to complete their cycle and become organic matter. Grass takes at least six months to double its mass. Weeds can double their mass in 10 days or less. Grass when it is grazed uses up fertility and minerals in the top layer of soil. Pasture grass therefore has a negative impact on fertility. Pasture grass gets eaten but also has a much smaller green surface area, so it is not capable of manufacturing as much organic matter through photosynthesis. The quantity of green surface area is an accurate measure of the productivity of the landscape i.e. the production of energy by photosynthesis. The key lesson here is that weeds are more beneficial than grass.

By photosynthesising the energy they draw from the sun, plants manufacture the compounds on which everything else depends. Yet mankind still keeps ploughing plants up and killing them with herbicides and destroying forests.

5. Soil Tests.

Soil tests are the key to good management of your major asset and should be done at least annually. There is no more effective way to find out if your soil is healthy and balanced and to have the desired pH of 6.4 than to do a soil test. To correct deficiencies and imbalances find a company that will make up a "prescription blend" fertiliser. This is a fertiliser made up specially for your soil in the right quantities to match your soil test and for the size of the area of land you are growing on. It is the quickest way to adjust your soil pH to the magic 6.4. It is normally only required once for the area you did the soil test for. Afterwards you normally only need a maintenance program to keep the fertility level up, to replace the elements lost through leaching and from what the plants take out. We use Nutri-Tech Solutions based in Yandina on Queensland's Sunshine Coast.

6. Fertigation.

A fertigation system is a method of putting liquid fertilisers into the irrigation system so that the fertilisers are sprayed on the plants through your sprinklers. They are a huge timesaver and help ensure you do the frequent fertilising program that makes such a difference to the health of your crops. There is no way I could fertilise monthly without a fertigation system. They cost about $2000 to install. Regular fertigation is essential for success in your venture. Fertigation is discussed in more detail later. Putting in fertigation systems (one for fruit trees and one for vegetables) were two of my best investments.

7. Foliar sprays.

Foliar sprays are probably the most effective way to fertilise your plants and ensure good yields. They are basically spraying liquid fertilisers in a water solution onto the leaves of your plants. The leaves can take up the nutrients immediately whereas fertiliser on the soil can take months to be absorbed through the root system. However you need to do both because in the long run plants need their root system to survive and thrive.

Foliar sprays supplement your fertigation program and have the special value of being able to correct any deficiencies or problems detected in nutrient uptake by plants. Nutrients taken up by through leaves have an immediate impact, whereas fertilising through the soil and roots can take months.

It is recommended you do foliar sprays every three months with fruit trees and monthly with vegetables. There is one foliar spray that benefits every plant and that is to do one with boron before flowering. With fruit trees this generally means anytime from late summer through till late winter.

With climate changes becoming more common we have recently seen flowering on avocados start earlier than ever. Here on Tamborine Mountain "normal" flowering on avocados has been around late September i.e. Spring. In 2008 many of our avocado trees on Mount Tamborine were in flower at the start of July (following a very mild winter up till then). The process of flowering leading up to the actual flowers appearing takes many months, so to be on the safe side for avocados you probably need to put a boron foliar spray in January/February and again in the May/July timeframe.

If you decide to do the spraying yourself, you should not skimp on the spray equipment. You need sprays that create a cloud of fine mist to ensure all the leaves are affected, particularly their undersides where the stomates are. To do this you need a fairly powerful spray unit to reach the leaves that are higher up. The more powerful units also save a lot of time. You can of course pay a contractor to do this job for you to save the capital outlay.

8. Leaf tests.

In the same way that you need soil tests to tell you precisely what is in the soil so that you know what to fertilise with, you need leaf tests to tell you what nutrients are being taken up in the plants so that you know what to put in the foliar sprays.

I take 20 to 40 leaves from the plants to be tested and send them in to a laboratory, such as Environmental Analysis Ltd.(EAL) at Lismore in NSW. The laboratory sends a copy of the results to Nutri-Tech who send a report analysing the results with recommendations for what to put in a foliar spray. Leaf tests cost lower than a soil test, which costs about $120. Currently the cost for a leaf test is $49.50.

When doing the spray, I normally include some fulvic acid and "cloak spray oil" which helps the spray to stay on the leaves and not get washed off. The impact is almost immediate.

9. Windbreaks.

Wind is often a much underrated factor in the growth of plants and trees. I had an early lesson of the impact of wind when I visited the avocado orchard of a friend. He had planted two groups of avocados, those out in the open and those under netting. I went back a year later and was amazed to see that the trees under the netting had grown to be almost twice the size of those exposed to the wind. Wind breaks are covered in detail in chapter 6.

10. Watering.

Water is one of the most important factors in successful farming. It is important in its ready availability, its quality, the frequency of watering, the amount of water needed, the timing of when to water, the variety of irrigation systems, the type of sprinklers, above or below ground pipes and so on. A whole book could be written about the significance of water in farming and how to use it effectively.

The hotter the climate the more you should consider underground irrigation i.e. no sprinklers above ground because of the huge loss of water due to evaporation. I deal in more detail with these issues in Chapter 16.

11. Trace elements.

Our soil analysis laboratory provides information on about 20 major elements. This means I do not have any information on the remaining 72 elements that make up the soil. My problem is: "What do I do about ensuring they are present in the soil?" The twenty I have information about are important, but many possibly all of the others are also important for maintaining healthy soils, healthy plants and healthy humans. They are needed in only very small amounts but to ensure they are present you need to identify a source of trace elements and the best source is the sea (where all the elements have been washed down by rivers for centuries). Seaweed fertilisers are great for the maintenance of soil fertility as they contain most if not all of the trace elements. Nutri-Tech ensures that most of their fertilisers have trace elements.

Trace elements are necessary for plants that are to be nutrient-rich and healthy. This important subject is covered in chapters 13 & 16 and Appendix A. The key fact to remember is that plants cannot manufacture elements that are not in their soil, and the elements that make up the soil are not spread evenly throughout the planet. Many elements are quite rare. Selenium which has been identified as an important mineral in human health is almost non-existent in Australia's soils and most of Africa. Most of Africa has high incidences of cancer and HIV-AIDS, but Senegal in West Africa has almost no incidence of cancer or AIDS. It is interesting that Senegal happens to have huge quantities of selenium in its soil. (references 2, 3 & 4). A CSIRO study in South Australia could find no traces of selenium in vegetables, not even one part in a billion.

12. Carbon.

Carbon is the essential element for life and the basic measure of fertility.
Sustainable farming needs soil carbon. It is the essential element and key ingredient for success and profitability.

Plants extract carbon from the air (from CO_2) and from the soil. Organic matter breaks down into over 50% carbon. This is why compost and mulch are so important. They keep up the levels of carbon available to plants. You can give the soil a carbon boost by adding carbon-rich fertilisers such as Nutri-Store 180 (which is a microbe-rich, high carbon, composted fertiliser) or humates (Huma-Tech Liquid Humus or Soluble Humate Granules). I obtain mine from Nutri-Tech Solutions.

13. Plan to have produce to sell on a regular basis.

Picking every week of the year, or at least once a month, has many benefits. It sets up a good routine, it helps the cashflow and it provides continuity of work. The regular production encourages you that all is well and helps to provide a monitor of your progress.

Weather can interfere with this plan. In the 2007/2008 summer we had over three months of almost continuous rain and 2008 has also been wetter than usual. Rain interrupts this steady work program. Having a dry packing shed is useful but you soon run out of packing jobs when you have not been able to plant or pick. We do of course pick in the rain at times but it is not pleasant. So one of our activities when it's wet is to make preserves –jams, marmalades, pickles, chutneys etc. for family use, but there is a limit to how much of this you can do. This is why I plan to install a shadehouse that keeps out the wind and rain, but not the sun, so that we have workplace that can be used in all weathers. I plan to grow tomatos in the shadehouse, as tomatos have a high demand and crops produce higher yields when protected from the wind and the temperature extremes.

14. Do not use herbicides or pesticides.

These toxic chemicals are made by man to kill living organisms, whether beneficial or harmful. You need all the help you can get from the billions of bacteria, microbes, fungi and worms that keep your soil alive and convert the soil's elements into plant food. So why kill them with expensive chemicals?

USA data shows that18% of all pesticides and 90% of fungicides are carcinogenic. These are going on our food. They also find their way into our streams and into the sea, affecting our water and wildlife. The World Health Organization reports that three million cases of pesticide poisoning occur every year, resulting in more than 250,000 deaths (reference 5).

Chapter 4
How I became a farmer.

"I have heard of the rainbows, of the stars, of the play of light upon the waves. These I would like to see. But far more than sight, I wish for my ears to be opened. The voice of a friend, the happy busy noises of community, the imagination of Mozart... Life without these is darker than blindness." –Helen Keller.

Computers had been my focus for over 30 years, working mostly in an office in London and Sydney. My wife had been in the education field for over 25 years, starting in a schoolroom teaching Geography and Economics at a girl's high school in Sydney, and then moving into business. Here we were in the country knowing no one and far away from Sydney and the business world.

I felt lost for a couple of years and totally out of my depth. Training was my motivation to move. I thought with my wife's experience in education and training and my experience from my last job where I ran a successful national training company that we would develop our own training business on Mt.Tamborine. I had grand ideas that we would soon be talking at seminars around Australia and even around the world.
That was why I agreed to "retire" in 1997 to a small village moving a long way from our friends in Sydney. Our project was not well thought out and did not take off. I decided to get to know people on the mountain and get immersed in the local community, joining many of the local clubs and organisations and playing bridge and golf.

To play golf well was one of my goals and our property is next to the local golf course. I eventually managed to break 100! One day I was playing golf with one of my new friends, Alec, and during the game he said to me: "You have a lovely property with lots of spare land why don't you grow something on it?" I asked him what would he do and he said I'd plant 300 avocado trees and in 3 years I'd have 300 avocados per tree and at $1 each I would have $90,000 income per year for the rest of my life.

This seemed a good idea! Shortly afterwards in 1998 I got a contractor to bring in a back hoe and had 300 holes dug-- 6 feet deep to check the depth of our soil and to see whether there was much rock. We had beautiful rich red soil full of earthworms, so I threw in some fertiliser and planted 300 trees. Unfortunately I forgot to ask Alec how to look after them and within a year I had lost 50 of them! Some I lost to hares who took a liking to young green succulent shoots and the rest I poisoned by putting a handful of boron around each of the plants. Someone had told me boron was an essential element but I didn't realise the soil needs only 2 or 3 parts per million and too much is very toxic.

Still I was ever the optimist and I thought the rest will come good and bring me a regular income from July to December, the picking season for most avocados on Mt.Tamborine.

I thought it would be good to have income all year round and after a little investigation I found that the passionfruit picking season is February to July, so I put in 200 poles and wires and planted 200 passionfruit vines to climb up and run along the wires. This time I spent a couple of hours talking to an experienced passionfruit grower about how to look after them so I was "set".

A year later they were looking good until one morning I went outside and found half of them had wilted and died overnight! We had had our first frost! We did get a crop of passionfruit in the second year but that turned out to be our only real crop as passionfruit unlike avocados live only 3 to 5 years. I did plant another 50 to replace some of the dead or dying ones, but passionfruit were not a big success story for me and I eventually gave up on them. The avocados started to bring in a little money and looked a lot more promising, and in the meantime my wife, Bev, had started growing vegetables as we wanted to be self sufficient in food and found also that there was a market for vegetables. We decided to grow cabbages, cauliflowers and broccoli. Cabbages did well and we filled our stationwagon with large cabbages and took them to the Brisbane market where we got 80 cents each for about 200—a grand total of $160 income (less the cost of seeds, fertilisers and transport). From this little exercise we decided to grow crops that were less bulky and less heavy and fetched a better price. We found garlic fits the bill although the only drawback is that it takes six months to grow, so we then tried carrots, beetroot and rhubarb all of which have a much shorter growing cycle. Rhubarb also has a magical property. It is one of a few vegetables that you can pick every 5 or 6 weeks from the same plant. After about a year you can divide up the crown and make several plants from the one, so it multiplies each year.

Soil tests are necessary for all farmers and anyone who really wants to grow plants and trees successfully. How else do you know what's in the ground? The soil is your greatest asset. I learned this from a talk by Graeme Sait, the CEO of Nutri-Tech Solutions Pty.Ltd., in 1999 when he spoke at a Tamborine Mountain Local Producers Association (TMLPA) meeting on Mt.Tamborine. I went and got my first soil test done and saw that we needed more calcium. I bought three bags from the local hardware store and spread it around my 250 avocado trees. Later I found that I really needed three tonnes, not three bags to make a significant difference.

The pH of my soil in 1999 was 5.4 and I needed to move it up to 6.4 and that takes a lot of calcium plus other elements as well. Even then it takes about a year to change the pH—the time it takes for the microbes and other wildlife in the soil to convert these natural elements into plant food.

I heard Graeme talk again in 2003 and this time I had had several years of experience of making mistakes plus I had become more interested in learning about farming. We got another soil test and this time we carried out the recommendation from NTS to get a prescription blend fertiliser to correct the imbalances. I got 3 tonnes this time rather than 3 bags and it was one of the best investments in the farm I have made. It took nearly a year to be absorbed and then the pH lifted into the magical 6.4 range. My avocados started to recover from five years of neglect and poor nutrition. Our avocado production has increased each year since we did this.

Even more significantly I understood much more of Graeme's talk this time and I went outdoors with renewed interest and much to my amazement plants appeared after putting in seeds in the ground and trees started to look more lush, green and healthy. One day I had this amazing insight where I saw what a miracle it is to plant a seed and see a plant emerge from the soil and I seemed to see how alive the soil really is. In that moment I changed from being a reluctant farmer who hated getting dirty to suddenly loving what I do and being proud to call myself a farmer.

This moment became a major turning point in my life. I stopped thinking of myself as a city businessman and started calling myself a country farmer. And even more important I started really living in the present, **the now**, seeing what is in front of my eyes instead of focussing too much on plans and goals, living in the future. I went into our house and said to Bev let's enrol in Graeme's Certificate of Sustainable Agriculture course run on the Sunshine Coast. We did and both went along to do the four day intensive course in March 2004 and stayed at a nice bed & breakfast place in Coolum near the sea. By now we knew how fast Graeme talked and how much information he provided--which can easily put us into overwhelm, so we got a set of seven video tapes of one of his seminars plus a workbook and we went through the seven tapes twice before we got to the course. The certificate course runs for four days and provides an amazing wealth of invaluable information on soil health, plant health and human health with good notes for ongoing reference.

Graeme highlights the strong analogy between soil health and human health. In particular, I got a strong lesson on microbes. How the soil is full of billions of microbes who help convert minerals into plant food and how the human body's gut (stomach and intestines) should also be full of microbes to convert our food into digestible form. These have been severely depleted in most industrialized countries, because to help shelflife most processed and packaged foods have preservatives that kill off the microbes. Preservatives kill off the microbes in our bodies too. Most of us need to supplement our intake with microbes. Good digestion is a consequence of a large population of microbes in our gut.

Graeme provided excellent morning and afternoon teas and lunches to demonstrate what we should be eating. His morning teas were really breakfast, and he provides a home-made muesli that is fermented with a "probiotic". He got us all to taste a dash of this liquid straight. His probiotic is called BioBubble and is a liquid full of beneficial microbes. The first morning I tried it, I liked it but my body did not respond kindly to this new unexpected input and I was sick for a day and a half and couldn't get out of bed unless it was to stagger to the bathroom with diarrhoea. Fortunately this was a one-off adjustment by my body getting used to having its microbes back and I have had no nasty side effects since, and after the course we took to making our own and when at home have it every day.

Following the course the productivity of our farm lifted dramatically and I strongly recommend that you consider doing it yourself (details are in Appendix E).

It still took 2 or 3 years of hard work to get our soil right and to start applying what we had learned, but as a demonstration of how far we have come from an inauspicious beginning, we won the Queensland Small Business Champion Award for fresh food in 2007 and in 2008 were one of four finalists in the "Australian Vegetable Grower of the Year" awards. On both occasions we celebrated by having a few days holiday and attended the awards nights. The second time we flew to Perth for the annual awards night and then went on to the magnificent Margaret River region in South West Australia and tasted their wines and food.

Up till now we have done very little advertising. We've not needed to, partly because our business has grown faster than we can keep up with the demand, but also because we have received an amazing amount of unsolicited free publicity. We have been featured in newspapers and magazines regularly, and have been on TV four times recently—twice already in 2008. The first time was on "Mercurio's Menu" shown on Channel 7 featuring Paul Mercurio of "Strictly Ballroom" fame. He has a programme featuring regional food in Australia. We were on the Gold Coast programme where he interviewed Bev, me and one of our woofers and obtained a range of vegetables from our garden that he cooked up at the nearby SongBirds restaurant. Then we were interviewed at our local farmers' market at the Greenshed for another Channel 7 TV programme called "Queensland Weekender".

We started a weekly farmers' market on Mount Tamborine and we sell most of our vegetables at this market which operates in a large Green Shed, that has become the local name of the market. The Green Shed also featured in the ABC's monthly magazine called "Delicious" in their August 2008 issue and we were in Queensland Country Life in June 2008 with a two page spread including four colour photos.

One of the reasons for all this media coverage is probably that the rising oil price and food prices have started to make more people aware of the value of local food. There has also been an increasing interest in organic food. The growth in a range of illnesses has resulted in a greater focus on health and food issues, witness the escalating number of TV programmes on food and cooking.

Demand has been growing so fast that instead of advertising, our own focus has been on increasing the number of suppliers to the GreenShed. In 2004/2005 we decided to place a small advertisement in our local paper, the Tamborine Times, to see if there was any interest in people wanting to learn how to grow their own food organically. One small advertisement generated 60 phone calls! So my wife put together a training course called "Growing Healthy" and we have now trained over 100 people on Tamborine Mountain. We have also run a course in Brisbane and recently one in Beechmont (40 minutes drive south).

As we both love training and meeting people, in December 2006 we put together a sketch of a training centre we would love to build on our property. We spoke to our local council about the possibility of building a residential training centre in January 2007 and in February we had a designer draw up some tentative plans for ten cabins plus a communal building that houses a commercial kitchen and a dining room that doubles as a training room. In March 2007 we lodged a development application and we finally received approval in February 2008.

Chapter 5
Choosing a property.

"I know a man who has a device for converting solar energy into food. Been doing it for years… It's called a farm." David Stenhouse.

1. Water. To grow plants, water is the most important factor when choosing a property. Without water, there is no food. To grow plants, water and sunshine are essential ingredients that are hard to create. If water is scarce, a simple strategy is to collect and recycle graywater i.e. the waste water containing soap and residues from dish and clothes washing, showers and baths or spas. This water can be reused on gardens. It is sometimes also possible to obtain recycled water from outside your own property. Irrigation is almost essential to be able to supply regular amounts of water to your crops all year, and for new farmers this is a whole new area of learning which I will discuss in more detail later. Questions you should ask are: Does the property have a bore, dam(s) or a stream? What is the annual rainfall and does it come all year or only in a few months?

2. Marketing of the crops you decide to grow is crucial to the success of your venture, so distance from markets is a factor as is the size of the market. Some produce and crops are affected by distance and time more than others. If you decide to add value to your produce by producing olive oil, for example, distance is not so important. If like us you promote your produce as "fresh and local", distance to your market is a major factor. Our farmers' market where we sell much of our produce is 3 km from our property.

3. Organic Matter in the soil is a key factor in the success of your venture. This can be rectified but is very costly for farms with big areas. It is worth doing a soil test to see what the level of organic matter is. One of our major Australian banks did a study to find what the best measure was they could use to determine whether a loan to a farmer would be a good decision for them. The solution they came up with was to check the level of organic matter. The higher the organic matter the greater the success factor for the farm.
50 or 60 years ago most western farmers averaged 5% organic matter. Today this average has dropped to nearly 1% as a result of the use of industrial farming techniques: irrigation, artificial man-made chemicals, fertilisers, herbicides and pesticides. All have contributed to killing off the "life" in the soil and contributed to erosion and salinization. The level of organic matter on our property is above 10%. One of the great benefits of organic farming is the regular use of mulch and compost which provides a home for the billions of microbes, bacteria, worms and fungi that nature provides in the soil to convert the elements into plant food. Organic matter generally breaks down into over 50% carbon in the topsoil. Carbon is the essential element for life.

4. Soil is a farm's primary asset. Things you need to check out are the depth, the rocks, the texture, the levels of organic matter and the cation exchange capacity (see next paragraph) of the soil. The best way to assess soil fertility is to do a soil test. Take some samples from several points wherever you are likely to grow anything and mix them up in a bucket and send the sample off to a testing laboratory. I use E.A.L. at Lismore in northern NSW but there are several others (see for example www.soilfoodweb.com.au run by Elaine Ingham).

What you are looking for is a good healthy balanced soil with as many minerals as possible. The soil test will tell you what major mineral deficiencies and excesses there are. The main indicators that are important to look for are the pH and organic matter. pH measures the acidity or alkalinity. The ideal soil has a pH of 6.4 and the lower the pH the bigger the cost of fixing it. If the soil has a pH of 7 or more it is also a problem as there is probably too much of some elements and it will need to be rebalanced.

Humus is special because it contains both positive and negative sites so it can store cations (positively charged nutrients) and anions (negatively charged nutrients). This storage capacity relates to both the storage of plant food and its fuel or energy needed for its plant growth.

The other key indicator is the cation exchange capacity (CEC) which refers to the nutrient storage capacity of the soil and is an excellent indicator of existing or potential soil fertility. The higher the CEC, the better the potential. All soils comprise varying percentages of sand, silt, clay and humus. Both clay and humus are negatively charged, which means that positively charged nutrients (cations) can attach to them. Most of the major nutrients are cations (except phosphate, sulphate & nitrate). (See reference 1).

Table of Cation Exchange Capacities (CECs).

Sand	2 to 3
Silt	5 to 7
Heavy clay	30 to 60
Humus	250
Humic acid	450
Fulvic acid	1450

Our own vegetable area tested in 2005 with a CEC of 40, but by 2007 it had risen to 84. Our avocado area had a CEC of only 19 in 2005 and in August 2006 had risen to 32. By December 2006 it had risen to 56. Our blueberry area tested at a CEC of 75 in 2008.

The five major cations are calcium, magnesium, potassium, sodium and hydrogen. The more calcium in the soil the higher the oxygen. The more magnesium the lower the oxygen. Like you and me, all soil micro-organisms need oxygen. Oxygen is also essential for root growth, water and nutrient uptake.

A good balanced soil analysis provides a percent base saturation figure for just the five major cations. This "ideal" soil balance approach was first developed by Professor William Albrecht, who has been called the "Father of Soil Science". He identified the most productive percentages and ratios between the major cations. This ideal is generally Calcium 68%, Magnesium 12%, Potassium 3 to 5%, Sodium .5 to 1.5%, Hydrogen 10% and other bases 5%. The trace elements are also important but are minimal in terms of how much of each element a soil will contain.

When I do a soil test Environmental Analysis Laboratory (EAL) sends a copy of the results to (NTS) who prepare a report called a Soil Therapy Report with graphs comparing the various levels of the main elements to the "ideal levels" and giving an analysis and recommendations on how to improve the soil fertility. NTS also provides an excellent service whereby they will make up a "prescription blend" fertiliser that corrects your soil balance and improves its pH. The quantity needed varies according to the size of the area you wish to grow in so you have to estimate that and put the figure in with your soil sample. We had only to put a prescription blend on once and within a year our soil was up around the required 6.4 pH level and this has major benefits. Our productivity increased and pests and diseases went away when we did this.

In summary, unless the property is seriously full of rocks, you can improve all soils. For vegetables particularly it's really only the top 12 inches (or 30 cm) that are really critical. If necessary, you can import the soil! For trees you need to consider more depth, the level of the water table and so on, but nature is very adaptable. It's true that if the soil is really poor (and you will **know** this from your soil test) it will cost you more to improve it. As an example of what can be done with poor soils, you have only to look at the area of the Australian wheat belt in south-west Australia. In this region the natural soil is severely deficient in most minerals yet with heavy application of fertilisers annually, it has become highly productive.

5. **Slope and aspect are important factors.** The steeper it is the harder it is to work the land and the bigger the risk of rolling a tractor. Slopes up to 10% are fine. The slope can also affect the ease of access for equipment and trailers and tire you out walking up it! The slope also affects drainage and water runoff.

The direction of the slope in relation to the sun's path is also important. If it slopes towards the north (in the southern hemisphere) you get more sunshine than if it slopes to the south, particularly with tall fruit trees.

6. **The sixth factor to consider is wind**. Many consider that wind is the primary factor that decides irrigation frequency and the amount of water needed by crops. Strong winds stop the growth of plants as they use their energy to protect themselves from the wind. Find out where the prevailing winds come from in your area. Then check if there are natural windbreaks already in place or if you have to plant them. Trees take 3 to 10 years to mature so the more trees you have at the outset the better, particularly around the

boundaries. Trees not only provide windbreaks but give you privacy, cut down the noise and can of course give you a crop. I talk more about the value of windbreaks later on, but bear in mind how valuable they are. Windbreaks improve the growth of your plants, minimise temperature extremes, encourage more rain and help stop erosion.

7. Other factors are of course the price and the house. Is the house, the property and the area somewhere you would be happy to live in and create as your home? What are the distances from nearest towns and cities? What is the transport like? Remember you need to transport your crops and to bring in supplies. What is the size? Is this suitable for the type of farming you envisage? As I will show you five acres of farming area plus space for buildings, dams, driveways, and trees is all you need to be profitable. Apart from the house, you will need some sheds for packing and shelter for animals and equipment. Flat space for a greenhouse or shadehouse would be a bonus, and you may need space for a coldroom (but these are not very big).

8. Finally the climate is important too, not that you can do a lot about it. Is it suitable for the crops you have in mind and are you personally comfortable with it? It can be too hot in many parts of Australia for apples, pears and stone fruit or not hot enough for crops like mangoes and bananas. Avocados do not grow in the extremes of Northern Territory and Tasmania or Canberra.

I was born in England and love the warmer drier climate in Australia after all the cold and drizzle, but the northern half of Australia gets uncomfortably hot for me.

As you can see there are many factors to take into account when you choose a property before you can get started.

Chapter 6
Farm layout.

"I believe that service-whether it is serving the community or your family or the people you love or whatever-is fundamental to what life is about."
Anita Roddick.

One recent visitor to our farm had just bought a small farm and after seeing what we had done made the comment our farm is laid out wrongly because "We have bananas at the high point of our property and avocados at the bottom." Bananas love water and avocados die if they get too much water.

Farm layout considerations.

On steep slopes more rain runs off instead of penetrating the soil. On sloping ground there is more moisture towards the bottom of the slope. Vegetables grow better with plenty of moisture and fruit trees generally do better with better drainage (bananas are an exception and seem to love water). Grow your vegetables on your flatter, lower areas.

If you do not have much flat land, make your vegetable rows perpendicular to and around the slope. Each row then forms a dam to catch the water and slow the erosion.

If the slopes are really steep, you may need to consider alternating the vegetable rows with a thicker crop such as corn, which also makes a windbreak.

With slopes another factor is sunlight and obviously slopes facing north (in the southern hemisphere) do best.

In deciding where to grow, you also need to take into account trees which in a row can also be a good windbreak, but can also stop the sunlight. Plants growing in too much shade will not do as well and too close to trees have to compete for nutrients and can get damage from falling branches (particularly gum trees).

Weatherproofing.

Frost and wind are elements of climate that can cause considerable damage to your farm.

Frost.

There are four steps you can take to minimize frost damage:

- plant more trees. On Tamborine Mountain it is readily observable that trees help reduce the risk and impact of frosts. Our property has thousands of trees and in the past 12 years we have had frosts in only four years and in three of those we had only one frost. Not far from us (about 4 km. away) a friend's property has had multiple frosts almost every winter and the frosts have taken a huge toll. Last year all their avocados fell in July (a fruit that starts growing in the previous October and is normally ready for picking around end July). Many of their avocado trees lost all their leaves and some died. This was in 2007 and the only year so far when our property had 3 frosts in the one year (all in the one week). However we saw very little impact and our avocado crop was better than 2006.

- humus rich soil
- regular appropriate foliar sprays
- ensure a highly mineralised soil so that your plants have a mineral rich sap.

A seaweed extract from Durvillea potatorum (as found in NTS' SeaChange) has been found to give excellent results in research by the Tasmanian Dept. of Agriculture.

Studies on grapes show that these four activities can give you a level of protection equivalent to a temperature that is six degrees higher than the temperature being expected.

Wind.

Wind can be very harmful, drying out the soil, causing erosion and sapping the energy of plants. It is very noticeable that plants do not grow as well on their windward side. If you have a prevailing wind direction, you will observe that the best growth takes place on the leeward side. Trees that are protected from the wind grow bigger and faster than those exposed directly to winds. Wind leaches heat and moisture from both crops and plants. Winds above a certain speed move soil particles, losing your soil, fertiliser, nutrients and organic matter.

A good protective action is to create windbreaks, by planting rows of trees for this purpose. Windbreaks prevent heat and water loss. They create a micro climate. All trees help. Trees and windbreaks help generate more rain and reduce soil erosion.

The height of windbreaks determines the protected area. A rough formula predicts windspeed reductions in an area of two to five times the height of the windbreak on the windward side and up to 30 times the height on the leeward side.

<u>Thirteen benefits of windbreaks (reference 1)</u>

1. Increase crop yields as well as hay and pasture yields
2. Reduce soil erosion
3. Trap winter moisture for spring crops
4. Reduce the amount of water and nutrients needed by crops
5. Moderate temperature and humidity
6. Reduce crop input costs
7. Provide valuable wildlife habitat, such as nesting places for birds, which feed on insects in the fields
8. Reduce heating costs 10 to 40% when placed around the farmstead
9. Provide tree products, nut products and suitable areas to grow high-dollar medicinal herbs like goldenseal, ginseng and brambles (such as blackberries and raspberries)
10. Produce firewood and fenceposts
11. Provide as much as 50% savings in feed costs for livestock
12. Reduce wind damage to crop plants
13. Spread snow evenly in a field

There is a book on permaculture that has some great diagrams and cartoons illustrating the difference that windbreaks make (see reference 2).

Weatherproofing your farming also means looking after your workers, your equipment and any animals. You need some sort of shed/garage for your tractor &/or rideon and tools. We built a custom-designed packing shed in 2005, which has turned out to be an excellent investment. In this we have a washing area to clean vegetables i.e. to remove the soil from potatoes, carrots, beetroot etc. and a large workroom where we pack and store boxes, crates and fertilisers—except chicken manure which has too strong a smell to work near. After several experiences with chicken manure, my wife suggested we create a special area remote from the house and packing shed to store chicken manure. We made a special new entrance to our property with its own gate near the road for trucks to drop off not only chicken manure but also loads of mulch, rock dust etc.

The packing shed is a good dry work area for when it is raining or when there are strong winds. When it is raining one of the problems we face is what do our workers do? The packing shed provides part of the answer and we also make preserves (jams, marmalades, chutneys and pickles from any surplus produce).

Another solution to working in bad weather (which we have not yet done ourselves) is to install a shadehouse (or greenhouse) that stops the rain. This helps (i) plants grow better by providing protection from wind; (ii) mitigates the extremes of temperature and (iii) provide another place for workers to work when it is raining. Shadehouses require irrigation but can be very productive and also provide another protection against pests and birds attacking your crops.

Greenhouses or shadehouses

Shadehouses generally increase productivity so are worth consideration. If your area is subject to occasional hailstorms, then a good shadehouse provides protection against damage and loss of crops.

In colder climates greenhouses protect against the cold and extend the growing season. In hotter climates shadehouses or moveable tunnels protect against too much heat.

Both greenhouses and shadehouses protect crops against the damaging effects of wind and provide a work area for when it rains. If you have a big enough operation to employ a workforce then you need jobs that can be done out of the rain.

In Australia about 20% of vegetables are grown undercover. These are mainly tomatoes, cucumbers, capsicums and lettuce. (reference 3).

Shadehouses are more permanent structures than netting, which is often used to stop birds and insects attacking fruit. On Tamborine Mountain some commercial growers of kiwi fruit use netting and crops like blueberries are often grown under netting. Netting may not stop the rain and unless it is put on poles may need to be removed to do the actual picking. A shadehouse structure is more substantial and could pay for itself as it needs less maintenance and will have a door so that you can walk and work inside. Some shadehouses take machinery inside.

Chapter 7
Getting started

"You can preach a better sermon with your life than with your lips."—Oliver Goldsmith.

This chapter starts the process of telling you how to earn over $100,000 p.a. in 5 years from 5 acres of cultivatable land, growing fruit and vegetables organically. Obviously, your financial goals might be totally different from this. You might be happy to make sufficient to cover the costs of running a small acreage, or you might like an additional 30 or 40 thousand dollars income per year. Whatever the goal, the principles are the same. The difference is in the time you will give to achieving your goal and the commitment you have.

We have been farming for ten years now and made lots of mistakes and started from a base where we knew very little about farming. We also started without having a clear goal and without plans to achieve it. This book will provide you with the key steps to achieve this financial goal in five years. Alternatively you may choose a different goal and I suggest other goals in chapter 20. This book will help you avoid the mistakes we made and get results quicker.

You can learn from our mistakes and do it in 5 years and earn progressively more along the way. You can start earning in the first year, in fact within the first four months. For early income which provides the cashflow you start by planting vegetables because the majority take only 3 to 4 months from planting to picking. Some do take longer, such as garlic and ginger, so I will highlight which to plant first. Fruit trees take generally at least 3 years to start producing, so they can wait a few months to be planted. However you need to start planning for fruit trees in the first six months because in the longterm you will make more money from fruit, so you need to decide which fruit trees you are going to work with, where you will plant them, how much land they will need and then to start preparing the soil. We will cover these steps in detail.

The first key question is what will you grow? I mention over 150 different crops to consider. As I will emphasise several times, variety improves your probability of success, so I strongly recommend you choose an absolute minimum of ten crops. In our case we grow about 60 different crops: ten different fruit, nearly 40 different vegetables, ten herbs and five nuts and berries. Some crops fall into more than one category e.g. some people call avocados a fruit, others call them vegetables. Some put blueberries under fruit, others under berries, but it does not of course matter which category you put them in. I put avocados and blueberries in the fruit classification because they have the needs of fruit trees from a growing point of view and also need more land devoted to them than vegetables.

Upfront you need to make some decisions about what you want to grow. With most crops you can change your mind later without too much hassle. All you will lose is the time and experience gained by working with a certain crop, as there is something different to learn about each crop. However the one decision that can cost you much in time and money is which fruit trees to grow because they take much longer to mature than vegetables and take more space. The rewards for your efforts with fruit trees are proportionately much greater. But start with some vegetables.

Whichever crop you choose, you will then be faced with further decisions: which variety to choose and for fruit trees also often which rootstock. I will discuss these decisions as we proceed.

How many varieties are there? This varies for each crop and the number used to be much greater than it is today but because big business has intruded on farming and six companies control most of the world's seeds, the varieties have decreased in number. Supermarket chains supply a major proportion of the western world's food and they prefer to focus on those varieties that have the longest shelf-life and that transport best. Square tomatoes with hard skins and a long shelf life would be chosen over all others. Taste is irrelevant.

In the past there were 5000 varieties of sweet potato grown in New Guinea alone, 3000 varieties of potatoes in Peru and a 100 years ago there were over 3000 varieties of apples. Today you are lucky if you can find ten different varieties of apples. The big variety meant different tastes to savour and each variety would be specific to a small area that had particular soils and climates which favoured them.

The principles to remember when deciding what to grow are (i) the more common the crop the greater the competition you will get, (ii) the less common it is the harder it will be to sell, and (iii) the more common the crop the lower the price.

As an example of the decision-making process I will use avocados, our first crop. You can follow my example, and choose your crop because someone tells you that it is a good thing to do. This is what I did. It is not the procedure I would recommend. I made this decision with very little knowledge and even less research. What I have subsequently learnt is that avocados can be grown successfully in most states of Australia, except for the extremes where it is too dry in Northern Territory and too cold in Tasmania, Victoria and ACT. There are 70 different species grown in Australia but by far the most common is Hass. On Tamborine Mountain where there are about 300 orchards, 80% of the trees are Hass. Fuerte are the next most common variety here and they have the advantage of being ready to pick two months earlier than Hass in our climate. They have a disadvantage that their skins are much thinner so they bruise easily and thus do not travel as well. Reed is another avocado type that has the advantage of extending the season and is ready to pick later than Hass. If you want to have avocados available for a large part of the year, grow 10% Fuerte to pick early, 80 % Hass to pick in the middle of the season and 10% Reeds to pick late. The next decision with avocados is what will you plant i.e. seeds or saplings, traditional saplings with one graft or clonal saplings with two grafts.

Growing from seed is not recommended for avocados because you will not get many fruit for seven or eight years, whereas grafted saplings start to have fruit in years two or three, although they take 5 to 8 years to fully mature into maximum production. They are normally grafted onto rootstock from a good quality tree. Some people learn to do this themselves. Most buy from a nursery. There are several avocado accredited nurseries in Australia (see Appendix D). Traditional avocado seedlings or saplings cost around $11 to $15 each and clonal seedlings are about $30. Clonals have been demonstrated to provide much stronger resistance to the main disease that affects avocados (called phytophera or rootrot—a common fungus that attaches to the roots and reduces the nutrient and water uptake and can be deadly if not treated).

You also have a choice of different avocado rootstocks. Again there is much research on which are better. I recommend Velvick rootstock. Apart from all these decisions there is also an investment cost. One hundred trees @ $30 each is $3000 for avocados alone. Fruit trees cost a lot more than vegetables to set up and take more space and time, but as I will demonstrate this investment really pays off. By comparison if you buy a few thousand carrot seeds you will spend maybe $300 to $400.

Blueberries came out on top on my first serious research into what are the most profitable crops to grow. You can sell 125g for $5.50 i.e. $44 per kg! (see photo). They also have the advantages of being small and light in weight and the blueberry grows on a bush that rarely reaches two metres in height and takes up much less space than say avocados or limes. So blueberries can be planted just a metre apart. I found that there is a nursery near Lismore in northern NSW called Mountain Blue Orchards that specialises in blueberries and has 70 acres of them. As in most crops there are quite a lot of varieties to choose from. In my first planting of 120 I tried six different varieties and asked for a spread in terms of time for picking, so that on Tamborine Mountain we could pick from September to December. After a year I had lost twenty and some others were looking sick. Fortunately in April 2008 Graeme Sait visited our property and we took him on a tour and he commented that some of my blueberries were stunted and had yellow leaves. Apparently blueberries normally grow in quite acidic soils, and we were growing them in a soil at 6.4 pH. Whilst all plants do OK at 6.4 there are a few minerals that plants can take up easier at lower pH levels. Graeme did a test on the leaves and found that sulphur and aluminium were lacking because of this pH phenomenon. I found there was a fertiliser product in our local hardware store called "Hydrangea **Blue**" which is basically aluminium sulphate, so I put this on my **blue**berries and lo and behold in a few weeks the leaves turned green and the plants started picking up. Apart from aluminium and sulphur, Graeme also suggested they were low in nitrogen, boron and potassium, so I made up a foliar spray as well with these five elements. This encouraged me to go ahead with our second planting and we picked up another 200 blueberry plants from Lismore in May and planted them in June 2008.

So to summarise, when deciding what to grow there are several key questions to be addressed. Which crop? Which variety? How many will I plant? To decide how many to plant, you will need to do a business plan taking into account costs and space requirements and I cover these later in the book. Do not forget that the climate of your region has a major impact on which crops will do well there. Research the crops that do well in your region. Talk to local growers.

Decisions like what to grow and how many to grow will have a big impact on your success, so take time to consider these choices. One of the big differences between conventional farms and what I recommend is the number of crops grown. There has been a trend in the past 60 years in the western world towards bigger farms growing only one or two crops, although in Australia I have seen some statistics that suggest each farm has three or four sources of income which is slightly better. But I am advocating a smaller farm growing a variety of crops. It would appear easier to go down the path of growing only 1, 2, 3 or 4 crops. This seems to focus your skills and be simpler, but here is why I choose variety.

The big reason is that it minimises the real risks that farmers face: 1. price fluctuations, 2. climate changes, 3. pests and diseases. The larger the crop the more you will risk attracting the pests and diseases specific to that crop. Furthermore getting a disease or virus can easily wipe out a major part of your income. We have seen this happen with rhubarb on the mountain which is a major crop here, and a virus has wiped out the whole crop twice for one farmer in recent times. Biodiversity is another weapon in defeating pests and diseases, but the primary weapon is to get the soil right (i.e. a balanced healthy soil with a pH of 6.4).

There are also other reasons. Supplying a variety of produce to the local farmers' market ensures a constant stream of income throughout the year. Marketing considerations mean that my recommended way to market encourages variety e.g. with our weekly farmers' market you need variety to encourage customers to come back regularly and to buy all their fruit and vegetable needs at the one place. You can control your own decisions on what to grow, but not other people's decisions. If five growers are selling, for example, broccoli and your only crop is broccoli, then you are unlikely to sell much. If you have three or four other crops to sell as well as broccoli, you will at least have some income for the week. Secondly for me it's much more interesting to have variety. You can learn more and have a much better choice of foods to eat. It would be pretty uninteresting to grow just garlic which fetches a good price, when you can grow bananas, avocados, paw paws, blueberries, tomatoes, boysenberries etc. Furthermore variety in our diet has been shown to improve health and longevity.

Later on I will show you our revenues for many crops and discuss the profitability of the crops we have chosen to grow.

Chapter 8
What are the principles to use to decide what to grow?

"You cannot control the world outside, but you can choose what you will bring into yourself. If you do not see anything of value in your life begin by growing a flower, vegetable or tree every day until it becomes a habit. You will discover much of value." **Geoff Buckley.**

What I have done in this chapter is draw out some criteria common to all decisions on what to grow. Then in the next four chapters I deal with each of the types of crops that need to be considered in four categories: vegetables, fruit, herbs & spices and nuts & berries. I have taken five chapters to cover this one decision thoroughly because it is fundamental to achieving success in your venture. The following are some of the criteria that are common to all produce when making these important decisions: diversity, profitability, nutrient or health value and climatic suitability.

1.Diversity.

The reasons for diversity are many.
- protection against unexpected weather, which is increasingly important with climate change.
- protection against price fluctuations. As a grower you have no control over the market, or what is being brought to the market. This is why small growers sometimes make a loss. They grow one crop and want to sell it at the same time everyone else is selling. In situations of over supply, prices drop because supply exceeds demand. One year we saw a situation where there were major imports of avocados into Australia from New Zealand right at the peak selling time and prices crashed.
- protection against pests and diseases. The greater the variety of crops the less the attacks, e.g. on the mountain there has been a virus attacking rhubarb and some large growers had their crops wiped out; we had very little impact on our rhubarb. But if you were unfortunate enough to lose one crop and you grow 60 different crops the impact would obviously be small.

2.Profitability.

There are several factors that affect profitability. Some crops are much more profitable than others. Which are most profitable? This information is not generally readily available—it is one of the trade secrets. I can share with you what I have learned.

There are at least eight factors that affect profitability of the crops you choose to grow: growing time, space, durability, price, demand, packaging, labour input and the cost of nutrient replacement.

The growing time from planting to picking. This is shortest for vegetables and herbs and longest for fruit trees and nuts. Vegetables vary from two to nine months e.g. rhubarb has a cycle of 5 to 6 weeks between pickings, garlic takes six months and ginger nine months. Most vegetables take about 3 months. In the next chapters I will give you typical growing times as we deal with each crop. The growth cycle can also vary a little in summer and winter.

Space required is another consideration. This has two aspects e.g. avocado trees need to be spaced 4 to 6 metres apart but blueberry bushes only one metre apart. And you can plant hundreds of carrots in a short row. The second aspect is the bulk or weight of the fruit or vegetable when you pick it and transport it e.g. cabbages, water melons and pumpkins are heavy and bulky and take a lot of space, but limes, blueberries and peas are light and small. Generally I have found that smaller, lighter crops are much better but they can take more time to pick. We had an early lesson when we grew cabbages and had to transport hundreds to Brisbane. They take a lot of space and fetch very low prices. By comparison garlic is small and fetches a high price. Garlic however takes more space to grow than many other crops e.g. carrots.

Fragility or durability is another factor e.g. Hass avocados are popular with supermarkets because they have a thick skin and travel well without bruising. Fuerte avocados may taste better but have thin skins and bruise easily, which means that there is a much lower demand for them and they fetch a lower price.

Price and demand are of course vital. Many fruit and vegetables are not very popular or well known, such as kohl rabi or mangosteens, but everyone buys potatoes, tomatoes and carrots. In the next four chapters I highlight which crops are popular and which do not sell well generally.

What is the pricing principle to remember? The more common the crop the greater the competition and the lower the price, whereas the less common the crop is the harder it will be to sell. Some less common crops fetch good prices but you will need to do your home work on marketing them. If you can find a niche market, you can do well with some of the more exotic crops, but generally it is much easier to sell what is in high demand.

You need to keep an eye on the prices of your competition so that you do not price yourself out of the market. It is a good practice to make a note of the prices in supermarkets and greengrocers when shopping; some wholesale agents will fax you their price lists; they generally review their prices of fresh food every week.

It is not always easy to compare one product with another, as some are sold by weight, some are sold individually (such as pineapples, paw paws, cauliflowers and cabbages), some are sold by the bunch (like herbs) and some are packaged in punnets (such as cherry tomatoes and strawberries).

Packaging can be an extra factor as it adds to your costs e.g. selling avocados to wholesalers can require the use of special trays and plastic "plicks" to put each avocado into and they also have to be carefully graded by size e.g. big avocados might be 16 to the tray, whereas small avocados might be 26 to the tray. This all adds to the cost and time and is another reason to sell direct to the consumer through farmers' markets.

Labour input.

The time it takes to plant, pick, wash and pack varies greatly from one crop to another. For example, garlic is time consuming to plant and pick, but far more carrots can be planted easily and quickly and carrots are also much easier to pick. Broccoli and cauliflowers are easy to pick whereas peas and beans take a long time to pick.

Cost of nutrient replacement.

It is far more expensive to grow garlic and ginger, than lettuce because of the amount of replacement fertiliser you need. Lettuce is mainly water. There is therefore both a plus and a minus impact of produce with high nutrient content; the plus is the added health value to humans, the minus is the cost of replacing the nutrients lost from the soil. You can see which items have this impact from studying the next point.

3.Health value.

Another important consideration when growing food is the value to you and your customers' health. This can also be significant from a marketing/selling point of view. The best measure of the health value of fresh food is what is called the ORAC figure. ORAC stands for Oxygen Radical Absorbance Capacity. The USA Dept. of Agriculture (DOA) tests the ability of foods and other compounds to subdue oxygen free radicals. This enabled the DOA to determine each compound's antioxidant capability. The ORAC values shown in the following table reflect these findings. By getting a generous mix of the top choices in your diet, you can maximize your body's ability to defend itself against cancer and other diseases.

Pigments in fruit and vegetables are phytochemicals—compounds that impart flavours and scents, fight off insects, defend against solar radiation and discourage harmful microbes. These pigments are antioxidants which fight free radical damage in both plants and people. James Duke, the author of a huge DOA database on phytochemicals, states: "We evolved with these phytochemicals and the body needs them desperately. Cancer in many cases is a deficiency of phytochemicals and the same applies to heart disease and the degeneration associated with ageing." Sunburn and wrinkles are a consequence of free radical damage and cataracts involve free radical attacks to the lens of the eye.

Blueberries are high in ORAC value. Anthocyanins are the key blue/purple pigment, but blueberries contain nearly 100 powerful phytochemicals. Duke's database states that blueberries are "analgesic, antibacterial, anti-cancer, anti-inflammatory, antioxidant, anti-sunburn, anti-ulcer and immunostimulants."

A fairly complete list of ORAC Values is set out in Appendix B. Sample ones are:

1. Vegetables.
Kale   1770
Spinach   1260
Broccoli   890
Carrots   207

2. Fruit.
Plums   949
Oranges   750
Bananas   210
Apples   207

3. Herbs & spices.
Cinnamon 267,536

4. Nuts & berries.
Blueberries 2400
Strawberries 1540

4. Climatic suitability.

Climate clearly plays a role in choosing your crops. Mangoes and avocados would not do well in the cold weather of Tasmania and Canberra. Some crops do like cold weather e.g. apples and pears and stone fruit (peaches, nectarines, cherries, plums etc.) all do well in places like Orange in NSW and Stanthorpe in Queensland which are two of the coldest places in Australia. Mangoes and bananas do well in the hot weather of northern Queensland. With the use of greenhouses and shadehouses you can extend the regions that will grow a crop by protecting against the extremes of hot and cold. Trees and windbreaks also moderate the climate. If you do not have access to water for irrigation, then rainfall will make a significant difference.

Chapter 9
Deciding what vegetables to grow

"Progress is all very well, but it has gone on long enough."—Ogden Nash.

It is time to make some decisions. Vegetables will generate some cash within three to four months of planting.

What are the options?

There are 44 common vegetables, but your choice in any one area is restricted by factors such as climate, labour availability and demand:

Artichoke	Kohl Rabi
Asparagus	Okra
Baby spinach	Onions
Beans	Parsnips
Beetroot	Pumpkins
Bok Choy	Potatoes
Broad beans	Radish
Broccoli	Rocket
Brussel sprouts	Silver Beet
Buttonsquash	Spinach
Cabbages	Spring Onions
Carrots	Swede
Cauliflower	Sweet potato
Celery	Tat soi
Capsicum	Tomatoes
Corn	Turnip
Cucumber	Silver beet
Eggplant	Snow peas
Endive	Sugar snap peas
Leeks	Rhubarb
Lettuce	Yakon*
Kale	Zucchini

* Yakon is a South American root vegetable that is rather like a sweet potato, but is so sweet you can eat it raw and it tastes a bit like an apple.

What are the top selling vegetables?

Market acceptance is one of the key factors to take into account. The vegetables that sell best at our local farmers' market in order are:

Potatoes, Tomatoes, Carrots, Onions, Lettuce, Pumpkins, Runner Beans, Sweet Corn, Peas, Cauliflowers, Broccoli, Red capsicum and Asparagus.

How long does it take from planting to picking?

In Summer it takes slightly less time than in Winter. Some crops can be planted and harvested several times a year, e.g. rhubarb can be harvested 4 to 8 times a year on Tamborine Mountain. Some crops take as much as 9 months between planting and harvesting. We plant ginger in September and pick in May. Garlic also takes six or seven months. We plant in Feb/Mar and pick in September. The time we switch our plant selection from summer crops to winter crops is at the end of February through March (when there is a drop in night time temperatures and the early mornings turn chilly). We start planting summer crops in September. This timing will vary from place to place depending on the latitude, aspect and altitude.

Set out below are typical times taken from planting (in weeks):

7 zucchini, radish, coriander
8 silverbeet
9 lettuce, beetroot, kale, cabbage
10 basil, snow peas, broccoli
11 carrots, parsley
14 celery
20 onions & potatoes
22 caulis

Crop rotation

Some farmers leave land unplanted for a year and maybe grow some crop like clover to plough in to help refertilise the soil. We think this is unnecessary if you continually fertilise and mulch. We have about sixty rows in our vegetable area and plant once a month, in the two weeks leading up to the full moon. After harvesting we mulch and fertilise. When we have completed harvesting a crop, we replant that row with a different crop. We mulch between the rows each time we plant and then mulch around the plants when they are big enough.

Nutrient Value?

Kale is top of the vegetables list and spinach and broccoli score well (see Table of ORAC values in Appendix B).

Seeds v. Seedlings

Hybrid seeds tend to be more expensive and cannot be saved for future crops. They are of course controlled by the big six food multinational companies. Monsanto, Syngenta, Dupont, Mitsui, Aventis and Dow who now control 98% of the world's seed sales. They invest heavily in research the purpose of which is to increase food production capacity. This comes at a cost because their seeds have may have built in terminator genes and are often genetically modified.

"Seed saving has always been a vital part of the agricultural cycle. Without saved seeds there is nothing to sow and nothing to harvest. This is a fundamental law of annual plant cultivation. What has traditionally been viewed as a natural right is being transferred by globalizing corporate interests into a legally granted or denied privilege. Seed saving is a skill we can and must reclaim." (reference 1).

There is a groundswell towards "heirloom varieties" and a movement towards protecting seeds. Where we buy seeds is set out in Appendix C. There is an Australian Seed Savers Network (reference 2). In the USA the Seed Savers Exchange publishes an annual yearbook with over 12,000 the seed varieties cultivated by members (www.seedsavers.org).
For an excellent book on seed saving techniques see reference 3.

A few other sources of organic/heirloom seeds are to be found on the following websites:
www.seedsofchange.com
www.greenpeople.org/seeds.htm
www.groworganic.com

Not all vegetable crops are propagated by seed. The major crops that are not are potatoes (tubers), sweet potatoes (tubers) and garlic (cloves). Some others are artichokes (root sections), asparagus (crowns), rhubarb (crowns) and various oriental or asian vegetables.

Vegetables are quick-growing high return crops. Most require more potassium than any other nutrient, followed by nitrogen. The exception to this is peas and beans that are high in protein of which nitrogen is a major component. Celery takes up a lot of calcium. You can produce very high yields of celery and tomatoes in a small space so these can be very profitable.

Some vegetables are established by direct seeding. Others are best transplanted, after being initially established in boxes.

Those that are generally seeded directly into the ground are carrots, onions, beetroot, turnips, shallots, parsnips, sweet corn, pumpkins, peas and beans. Whereas cabbage, cauliflowers, broccoli, silver beet, celery and leeks are generally transplanted.

Successful establishment by direct seeding depends on two factors:

1. the ability to place the seed in the right place and quantities—either by hand or machine.
2. optimal seedbed preparation.

Our seedbed preparation involves mounding, loosening the soil with a garden fork and putting in chicken manure, gypsum and rock dust.

The advantages of transplanting are:
1. It shortens the growing season thus permitting more efficient use of the vegetable growing land.
2. Improves crop uniformity.
3. Permits more accurate prediction of harvesting dates.
4. Eliminates the time & cost of thinning.
5. Permits greater control of early growth and development.
6. Allows culling of seedlings showing poor growth and vigour.

Disadvantages of transplants compared to direct seeding:
1. Transplant shock.
2. Cost per seedling may be too great where a large number are required; extra labour in establishing the crop.
3. Can damage the tap root.
4. If transplants are delayed from planting out (due to adverse weather or delays in land preparation), yields may decline and maturity is delayed.

Vegetables response to transplanting

Good for transplanting	Need care	Not good for transplanting
Broccoli	Beetroot	Beans
Brussel sprouts	Capsicum	Carrots
Cabbage	Celery	Corn
Cauliflower	Eggplant	Cucumbers
Silver beet	Leeks	Parsnips
Tomato	Onion	Peas
		Pumpkins
		Radish
		Turnips

Pricing of vegetables is generally fairly uniform. I find there are two categories:

1. Peas, beans, carrots, beetroot, asparagus, broccoli, celery, cauliflower, potatoes, spinach or silver beet, tomatoes, sweet potatoes, pumpkins and zucchini all sell well.
2. Brussel sprouts, asian vegetables, kohl rabi, leeks, eggplant, okra, radish, corn, button squash, turnips, swede, and yacon all tend to sell far less and are harder to sell.

Lettuce sell well, but there tends to be plenty around at the same time and often they do not fetch as good a price as the first category.

The vegetables that consistently fetch the best price are celery and cauliflowers.

So which are the most profitable vegetables?

It is not obvious which vegetables are most profitable as many factors come into play. We have a standard row length which is about 20 metres. Broccoli yields about $300 + per row. Carrots yield about $600+ per row. Kale gets $330 per row but you can pick at least four times from the same row, so is highly profitable if you can sell it. Fortunately for us we found a large customer who loves kale, because our local farmers' market sells only a few bunches a week. Peas and beans both take a long time to pick but both get several picks; peas fetch a better price than beans, but beans are more prolific.

Based on the factors of pricing, time to plant and pick, ease of preparation (washing and packing) and nutrient value, there are five categories (A is the most profitable to E which is less profitable):

A. Carrots, Rhubarb, Kale and Silver Beet are excellent.
B. Peas, beetroot, broccoli, cauliflowers, potatoes and spinach do well on all factors.
C. Celery, beans and tomatoes are close behind.
D. Sweet potatoes do better than most of the others.
E. Cabbage, leeks and spring onions.

Some other factors that affect profitability are density of planting and the ability to pick several times from the one plant. On this latter factor rhubarb scores best, followed by kale and silver beet. Peas and beans also pick several times. In addition, two other considerations are whether they are easy to grow (like silver beet) and easy to pick and pack. Some crops take a lot more labour e.g. in small quantities carrots and potatoes need washing which can be time consuming. When grown in bigger volume you can automate both the picking and packing and washing of carrots and potatoes e.g. with air blasters.

Your profitability can also be affected by what other growers do. This is the competition factor and locally if two or three other growers focus on growing large quantities of one particular crop this can adversely affect your sales.

What do we grow?

We have grown 42 of the 44 vegetables listed above (the only exceptions being artichokes and yacon, both of which are grown by other growers on the mountain).
We have the greatest area devoted to growing rhubarb (see photos 11 to 14). Carrots, beetroot and kale are next in importance (see photos 27, 36 and 38).

Chapter 10.
Deciding which fruit to grow.

"When one is willing and eager, the gods join in."—Aeschylus.

What are the options?

There are 50 main fruit to choose from and of these 25 are the most common (marked with an asterisk *):

Apples*	Mangosteen
Apricots*	Monsterio deliciosa
Avocados*	Nectarines*
Babacos	Olives*
Bananas*	Oranges*
Cherries*	Papaya or Paw paw*
Choko	Passionfruit*
Citron	Peaches*
Custard Apples*	Pears*
Damson	Persimmons
Figs*	Pineapples*
Grapes*	Plums*
Grapefruit*	Pomegranate
Greengage	Pummelo
Guavas	Quince
Honeydew melon	Rambutans
Jaboticaba	Rock melon*
Kiwi fruit*	Rosehips
Kumquats	Rosellas
Lemons*	Sapote
Limes*	Starfruit
Loquats	Tamarillos
Lychees	Tangelos
Mandarins*	Tangerines
Mangos*	Water melons*

What are the top selling fruit?

In Australia the top sellers are grapes, apples, oranges and bananas.

In Queensland the list is slightly different: bananas, strawberries, mandarins, avocados and mangoes.

How long does it take from planting to picking?

For trees to be mature enough to bear fruit takes generally two to four years, so you have to be much more patient than with vegetables. But like a lot of things in life, patience is rewarded because the payback is higher and once established most trees live for 15 to 100 years.

Avocados (grown from seedlings grafted onto established rootstock) can produce a few fruit in year two and generally the number of avocados per tree increases substantially each year as the tree grows until they reach their peak in about years 8 to 10. There are avocado trees on Tamborine Mountain that are over 50 years old, still producing large quantities.

Macadamias take longer than avocados before they produce nuts. I planted 25 in 2005 and am still waiting. We planted one trial tree 12 years ago and it has produced some, even though I did not look after it properly until recently.

Olives can take eight years to produce their first crop. Blueberries take at least 5 years to fully mature but start getting a few fruit after one year.

Passionfruit grow on vines, like grapes, and are expensive to establish as they need posts and wires and have a remarkably short life—only 3 to 5 years. We got our best crop in year two.

Health value

The health value of fruits is shown in Appendix B. Mangosteen, an exotic fruit that has no relationship to mangoes, scores very highly. Plums, oranges, grapes, grapefruit and kiwi fruit also score well.

Prices

The fruit that fetch the best prices are pineapples, cherries, mangoes, paw paws and rock melons.

Seeds or seedlings?

Whilst you can plant fruit trees from their seeds, it appears they take much longer to fruit. It is generally recommended you plant grafted seedlings or saplings that have been well established. In view of their longterm value it is well worth buying good quality seedlings. We planted one avocado from seed and it took about 9 years before it produced an avocado and even afterwards did not produce as many as a grafted seedling does in years two and three. There are only four accredited nurseries in Australia for avocados (see Appendix D). We have successfully planted tamarillos from the seeds in the fruit and have had less success planting cuttings.

What fruit do we grow?

We grow avocados, blueberries, paw paw, bananas, most citrus, custard apples, mulberries, tamarillos, finger limes and guavas (see photos 19, 20, 23, 24, 25, 29 & 30). We tried passionfruit but gave them away, when the first frost killed half of them overnight and we realised they had such a short life compared to most fruit trees.

Which fruit are most profitable?

Blueberries are one of the best crops as we saw in chapter 7. Their downside is the labour involved in picking them. They score well on health value, and they have the advantage of taking less space than other fruit trees. This latter factor is significant when compared to many fruit trees such as avocados which require four to six metres spacing between each one. Blueberries are also reported to live for up to sixty years (see reference 1).

In chapter 23 I show you some revenue projections for several fruit and from these, limes appear to be more profitable than avocados. This is partly because big business has got involved with avocados in recent years plus imports from New Zealand are contributing to dropping prices. Limes are more of a niche market, where there is a strong ongoing demand from bars and restaurants.

Apples and bananas always sell well and one of the less common fruit, custard apples, is surprisingly popular at our local market.

Mandarins are quite good sellers. Grapefruit and lemons do not sell well.

With fruit trees, unless you have a lot of land, you need to be more selective than with vegetables. It is easy to change from one vegetable to another, but with fruit trees your decisions have a much greater impact in terms of the size of your investment. This is because you need to dedicate more time, space and money to establish a longterm income than with vegetables. It is also worth spending more time researching fruit trees and developing expertise in the ones you select.

Chapter 11
Herbs and Spices

"The doctors of today prescribe medicines of which they know very little, to cure diseases of which they know less, in human beings of whom they know almost nothing."—Voltaire.

Herbs have been used since ancient times to add flavour to foods and for their medicinal qualities and wonderful fragrances. The difference between herbs and spices is often not known. Herbs are the leaves of a plant. Spices are produced from the other parts i.e. the flowers, seeds and roots. The "Encyclopedia of Herbs" by Deni Bown lists over one thousand herbs.

What are the main options?

The most common amount to 37 varieties, of which 17 are the most popular (marked with an asterisk).

Aloe vera*	Gingko
Angelica	Ginseng
Balm	Horseradish
Basil*	Hyssop
Bay	Lemon grass*
Bergamot	Lovage
Borage	Marjoram*
Camomile	Mint*
Caraway	Oregano*
Chervil	Parsley*
Chives*	Rosemary*
Comfrey	Sage*
Coriander*	Savoury
Cress	Sesame
Cumin	Tarragon*
Dill*	Thyme*
Fennel*	Turmeric
Garlic*	Watercress
Ginger*	

Herbs themselves are not big sellers in our market, but are a valuable nutritious addition to our foods. We find ten herbs provide a good coverage and encourage customers to come back. Herbal teas are an added value item for consideration. The top selling herbs are clearly garlic and ginger. The others in order of popularity are basil, mint, coriander, parsley, thyme and rosemary.

How long do they take from planting to picking?

There is a huge difference here between the different herbs.

Herb growing times:

Coriander	7 weeks
Chives	6 to 9 weeks
Dill	8 weeks
Basil	10 weeks
Parsley	10 to 11 weeks
Garlic	6 to 7 months
Ginger	9 months

Thyme and rosemary are a bit more tricky to grow. These are both perennials. You have to wait for at least a year before you start picking. Mint is also a perennial and generally grown by taking cuttings from the runners.

Garlic takes 6 to 7 months to grow and it fetches a good price; less time and a better price than ginger. It is the best priced herb, fetching $20 per kg. wholesale and $30 to $40 per kg. retail. The demand for local garlic is strong as more and more people are learning that imported garlic from China is often irradiated, so you cannot grow your own from it and the nutritional value is suspect. One enterprising grower in Tenterfield in NSW sells his by mail i.e. he posts orders because it is so light. This suggests you could do well advertising it on the Internet. Garlic however is time consuming to plant and to pick. It also gets planted 5cm apart and our strike rate is good if we get 50%. Digging up garlic is slow because it is small and you cannot pull it out. Garlic is very light and yields only about 3 kg per row i.e. a yield of less than $100 per row.

Ginger is next best in price for the herbs and fetches about $12 per kg. retail (see photo 38).

At our local farmers' market all other herbs are sold at $2 per small bunch.

Most herbs and spices are supposed to be good for your health but only cinnamon shows up on the ORAC table (although it is at the top).
Turmeric has been used to dissolve cancer sells. (reference 1). India has an extremely low Alzheimer's profile. In India they use turmeric the way we use salt and pepper. Turmeric

contains the special trace element yttrium. The special thing about yttrium is that injecting it gave a three-fold increase in the lifespan of the test animals (reference 2).Ginseng has all sorts of claims including improving endurance and longevity, but it itself takes endurance and time to grow, as the time to maturity from planting can be five to ten years.

Seeds, seedlings and roots.

Most herbs are grown from cuttings or by splitting the roots. The only herbs we grow from seeds are oregano, coriander, parsley and basil: these four are planted in seed boxes and then replanted as seedlings. Rosemary is grown from cuttings. Thyme you can split the roots or grow from cuttings. Chives can grow from seeds but generally you separate the clumps of little bulbs. Marjoram and mint you split the roots. Turmeric and galangal grow from bulbs or corms. Ginger breaks up into small nodules in June and is kept in potting mix in the dark with no water for three months is then planted out in September. With garlic, you divide the crowns and plant directly into the ground; but a lot of varieties will not regrow as they have either been subject to irradiation or have a 'terminator gene' so that you have to rebuy them from the chemical/seed companies.

Which do we grow?

We grow about 11 herbs, particularly garlic, ginger, coriander, parsley, basil and mint. We also have a large bay tree with a huge supply of bay leaves. Others we grow in small quantities are thyme, rosemary, chives, dill, turmeric and galangal. We also grow a little aloe vera, which is quite popular and has many uses.

Chapter 12
Nuts and Berries

"A garden is a friend you can visit anytime." Anon.

What are the options?

There are 14 common nuts and 7 of them are quite popular. There are 14 common berries and 5 of these are the most popular (marked with an asterisk).

Almonds*	Hickories
Beechnuts	Macadamias*
Brazil nuts	Peanuts*
Cashews*	Pecans*
Chestnuts	Pine nuts
Coffee beans*	Pistachios
Hazelnuts	Walnuts*

Bilberry	Gooseberry
Blackberries	Loganberry
Blackcurrants	Mulberry*
Blueberries*	Raspberry*
Boysenberries*	Redcurrants
Cranberries	Strawberries*
Elderberry	Wolfberry

An interesting nut or bean to consider is coffee. There is a coffee plantation on Mt.Tamborine and coffee is a big crop in northern NSW. A seven year old coffee tree can produce 7 kg p.a. If you decide to look into coffee, I strongly recommend a booklet put out by the DPI Qld. in 1995 called "Coffee growing in Australia". In Australia coffee is mainly grown in Queensland and NSW. World prices have fluctuated between $2 and $10 per kg. Production varies between 2 to 10 tonnes per hectare. Revenue typically $5 per kg x 5 tonnes or 5000 kg = $25,000 p.a. per hectare or 2.5 acres. This is quite a low return compared to many other crops.

What are the top selling nuts and berries?

In Queensland macadamias are the big sellers in the nut category. In the berries: blueberries and strawberries are the big sellers.

How long do they take to grow?

My experience with nuts is limited to macadamias and it takes them at least four years to produce their first crop.

Health or ORAC values

Whilst nuts are a great source of many minerals (e.g. brazil nuts are a significant source of selenium), none show up in the ORAC tables. Digestion of nuts can be a problem and fermenting nuts certainly eases this problem. For nut smoothies or mueslis it is recommended they be fermented overnight. Berries tend to dominate the ORAC tables with goji berries and blueberries being very high on the list. In fact most berries score well.

When to plant?

Our strawberries are planted in Autumn and fruit in October/November (see photos 10, 11 & 12). Commercial growers just north of Mt.Tamborine pick from June through to end October.

Pricing?

Macadamias are most popular. We sell unshelled macadamias in a bag for $3 at our farmers' market. They fetch more if they are shelled, but this adds quite a bit of labour content. Peanuts are the lowest price. Walnuts, almonds, cashews and macadamias are all quite expensive, generally $17 to $25 per kg. As mentioned blueberries fetch $44 per kg.(see photo 34) whereas organically grown strawberries retail generally between $3 and $5 a 250g punnet i.e. $12 to $20 per kg.

Seeds/seedlings/runners

With strawberries you generally plant runners each year and discard last year's plants. Boysenberries also produce runners like strawberries. Mulberries are grown from cuttings.

Macadamias are like avocados, grown from grafted seedlings and they become tall trees, taking many years to reach maturity.

Which do we grow?

Macadamias are the only nut we grow (see photos 23 & 24), but we grow four berries: blueberries, strawberries, boysenberries and mulberries. We have high hopes for blueberries and planted 120 in 2007 and another 200 in 2008. Their high price makes for good revenue projections (see chapter 23).

Which are the most profitable?

Clearly blueberries come well out in front. They get high prices, have high health value and take up less land than many other fruit trees, as they are really a bush and can be planted one metre apart. The risk with blueberries is that birds will eat them. A common solution to this problem is to put netting over them. Blueberries are unusual from a growing point of view in that they are commonly grown on more acidic soils. This is because they need aluminium, which is more readily taken up from the soil at lower pH levels. To overcome this you need to do a foliar spray on blueberries with aluminium sulphate.

Chapter 13
Soils, roots and leaves

"98% of all soil organisms live and work in a thin layer of earth a foot deep (or 30cm). They seize essential nitrogen and solar energy from the sky above and mine the soil for essential minerals that make plants grow."
The Bio-Gardener's Bible by Lee Fryer.

A. SOIL.

Due to the facts in the above quote we are mainly interested in the topsoil.

What is the soil made of?

Organic matter should make up at least 5% of your soil. Water can account for 15 to 25% of the weight of the soil. Oxygen and carbon dioxide account for 20 to 25% when measured by volume. The live part of the soil, all the trillions of micro-organisms, make up only .01% but this amazing number of species work 24 hours a day for you. The other 50% is primarily all the minerals and trace elements.

Organic matter comes in two categories. The first is raw and incompletely decomposed plant, bird and animal residues which are undergoing decay through the action of various microbes and worms. The second is humus which is decomposed organic matter and is brown or black in colour. Humus is rich in humic substances such as humic acid and fulvic acid, which help in making nutrients available to plants. These acids stimulate plant growth. Both fulvic and humic acid provide food for soil microbes and worms. Organic matter is typically over 50% carbon.

Soils are all different and they are changing all the time. Rain leaches away essential nutrients. Plants and trees, grass and weeds take minerals from the soil to grow and unless they are given back as organic matter when they die these nutrients are lost to the soil e.g. by animals eating them.

Soils typically have fair quantities of calcium, phosphate and potash (or potassium). Magnesium is usually high in dry areas and low in wet areas. Most soils also contain silica and aluminium. Some soils have quantities of iron, zinc and manganese and most have some trace elements. Not all of these elements are always available for plants to use.

Nature is diverse; there are red soils, black soils, brown earth and sandy and clay soils. All soils can be made fertile.

What are the essential minerals? I believe all of them are needed by plants and humans. Most are only needed in very small quantities but these are often essential to various processes. Most of plant growth comes from the air and water through photosynthesis (as we see in the next chapter) but the 5% that comes from minerals in the soil is absolutely critical to the health and productivity of plants.

Good soils need to be well aerated. Worms and ants help to aerate the soil. Water-logged, poorly aerated or very dense anaerobic soils are low in oxygen and suffocate roots so that they stop growing. Stunted roots cannot take up the nutrients required for plant growth. Many beneficial organisms die or become dormant in anaerobic soils. Good soils have a loose structure. Compaction squeezes out the air and destroys good structure.

Nitrogen (N) is used by plants more than any other plant foods. N is of course not an earth mineral but is supplied naturally for free by soil organisms that fix nitrogen and from the decomposition of organic matter. When supplied by man-made fertilisers nitrogen is very expensive. The carbon to nitrogen ratio is important and should be 10 or 11 to 1 (humus is generally 58% carbon).The other ratio of importance is the nitrogen: potassium ratio in the leaf which should be 1:1 (see reference 1).

Two other big plant food sources come from potassium and phosphorus. Phosphorus is the main source of energy for the plant. Equally important are calcium, magnesium and sulphur. Some plants need more calcium and magnesium than phosphorus.

The six major trace elements are iron, boron, copper, zinc, manganese and molybdenum.

There is a lot of important information about these minerals and their roles available. Most people will not remember all the data about them and the really important thing is to be able to find the information readily when needed so I have put the useful data about each of the main elements and trace elements into Appendix A. I refer to them frequently throughout the book; so you can refer to this Appendix and it becomes a useful summary.

B. ROOTS.

The bigger and healthier a plant's root system, the better the growth and yield of the top of the plant. Roots absorb water and available nutrients from the soil. Roots also release acidic substances, called exudates, which make nutrients in the soil available that were previously unavailable. Each type of plant puts out a different exudate. They release carbohydrates into the soil as food for the benefit of microorganisms.

The root zone relationship is based on the plant enticing the microbes by releasing root exudates, which are microbe food. As the plant grows, beneficial microbial colonies will also grow in size and this benefits the plant for its whole life. These benefits are increased nutrient availability and disease suppression.

Roots also interact with soil microorganisms e.g.fungi called mycorrhizae live in roots and help the roots to absorb water and nutrients; others trap nitrogen and extract unavailable nutrients from soil minerals and organic matter. Fungi help to increase the root's absorption surface area by as much as 100 times (reference 2).

Plants trap energy from the sun through the process of photosynthesis and they use this energy to produce food in the form of sugar. Sixty percent of a plant's energy goes to the roots and 50% of that energy is put out through the roots to attract and feed microbes. This means that 30% of the plant's total energy is dumped into the soil as exudates which feed soil microbes. **The more leaves plants have the more energy a plant will have to develop its root system**. Therefore it is important not to prune too much and when mowing around trees it is better not to mow the grass and weeds too low; in fact it is a good idea in orchards to mow alternate rows leaving one row higher. Then next time you mow do the other rows. By keeping the grass longer you will help the grass roots grow more as they will get more energy passed down to them.

Roots take up nutrients from the soil. Plants also provide food for the soil life through the root system. Only the tips of the root where the root hairs are, can take up nutrients. The faster a root grows the more root hairs it produces as a response to increased growth. This means the more energy a plant has, the more root hairs it develops and the more vigorous the growth. **The biggest demand for energy in the plant's life cycle is in the time of early root development. This is the time that determines the final yield from the plant, generally in the first seven days of a crop's life. For trees it is the first seven weeks.** Early nutrition yields good results. The plant seldom catches up so good phosphorus management at the start is vital because phosphorus is needed for energy. The amount of energy available determines the rate of growth.

Part of good management of phosphorus is to understand that the basic structure of clay is aluminium and silica and in clay soils phosphorus readily joins up with the aluminium, especially at low pHs, and is made unavailable to plants. Magnesium has a strong influence on phosphorus uptake, so a foliar spray with P and Mg is a good idea straight after planting. This is a backup to providing phosphorus in a slow release form in the soil. Another tool to help in the mineral uptake by roots is humic acid. It has been found that plant's are able absorb around 30% more through the roots (or the leaf) if applied fertiliser is combined with humic acid. Fulvic acid can further contribute to phosphate uptake. Fulvic acid is a mineral solubiliser, which can initiate the release of tied up reserves of phosphate.

Silica is the other key to mineral uptake in both roots and leaves. Si strengthens leaves and this increases photosynthesis which means more production of glucose and more sugar support for the root-zone microbes. The translocation of nutrients within the plant is enhanced by the presence of sufficient silica.

Roots help provide a continuous supply of water to plants to replace the loss of water through transpiration from the leaves. This upward flow of water is what carries the nutrients from the soil to all parts of the plant.

C. LEAVES.

Leaves are like solar panels, trapping the sun's energy. The more leaves, the more energy the plant has. This is why excessive pruning will reduce fruiting in the short term.

Photosynthesis is the single most important aspect of crop production. The green plant is the only food producer on earth, and all living creatures depend on it. 95% of crop weight is from photosynthesis; the other 5% comes from the soil. Silica increases photosynthesis by improving leaf strength and helping to reduce attack from pests.

Leaves have two way traffic, taking in CO_2 and water and transpiring oxygen and water. Leaves also take in nutrients through stomata which are the breathing pores of a plant mainly on the undersides of the leaves. Early morning or evening is the best time to spray as stomata are often closed in the middle of the day. Fulvic acid should be used with other nutrients in foliar sprays, as it increases the plant's ability to take up oxygen with an associated increase in growth. It also enhances the photosynthesis process and improves water storage.

Foliar feeding plants through their leaves with foliar sprays is many times more effective than feeding the roots with fertilisers. It is the quickest way to correct nutrient deficiencies. It is not a substitute for soil fertility; it is much more effective when used in conjunction with a balanced soil.

Leaves produce exudates like roots. These feed minute bacteria and fungi that live on leaves; their role is to reduce attack from pests and diseases. Other leaf bacteria reduce frost damage.

1. Bev & Geoff win small business champion award for fresh food (Photo by EventPix).

2. Our vegetable garden looking north

3. Plum Tree in blossom in August

4. Our new dam

5.   View south from our balcony

6. View from training centre building site

7. Seedling in boxes

8. Strawberry plants in pots

9. Strawberry plants in garden

10. Strawberries growing

11. Rhubarb growing.

12. Rhubarb in transit to packing shed (Bev on ride-on).

13. Rhubarb picked

14. Rhubarb packed

15. Lime trees: 100 ready to plant

16. Preparing soil for limes- cardboard and mulch

17. Limes just planted

18. Limes showing tall sprinkler between two trees

19. Twelve year old lime

20. Tamarillos

21. Macadamia-three years old

22. Macadamia twelve years old

23. Citrus-Lemons

24. Avocados in the mist

25. Avocado in flower with small sprinkler

26. Turnips

27. Kale

28. Woofers picking beans

30. Blueberries recently planted

31. Blueberries for sale at market

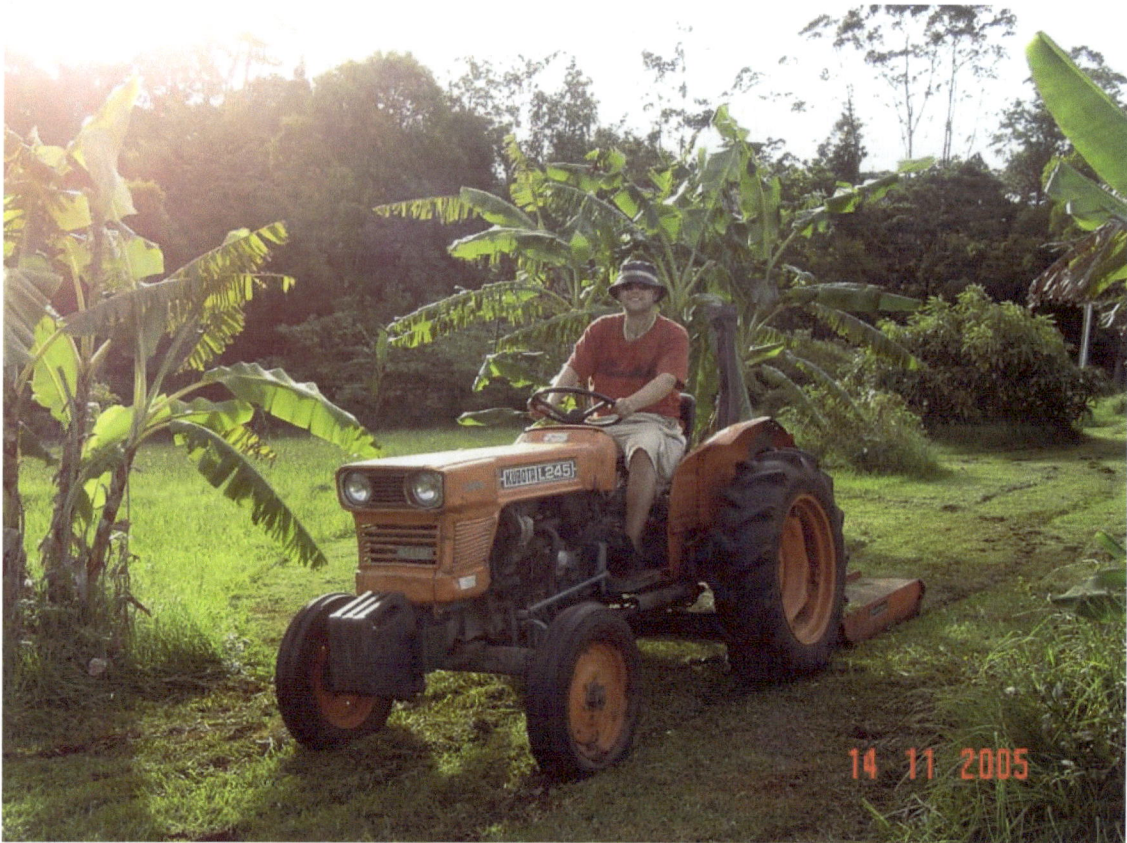
32. Tractor with slasher and woofer

33. Trailer with canopy- behind our stationwagon with packing shed in background

34.Fertigation unit

35.Foliar spray unit

36. Getting ready for market

37. Produce in trailer ready to go to market

38. Woofers picking avocados for market

39. Carrots and silver beet at our market

40. Customers at the market

41. Flowers at our market

41. Woofers washing ginger for market

42. Our ram with his harem

43. Rideon with trailer and sheep

44. Our first lamb

Chapter 14
How to grow

"The part can never be well unless the whole is well." –*Plato.*

You need only five things to grow any crop:

- Sunlight
- Air
- Water
- Organic matter (OM)
- A balanced soil.

The first three are usually free and nature provides two magical processes to make use of them and do much of the work. Your job is to focus on providing organic matter and a balanced soil. This is an ongoing job because rainfall, irrigation and wind leach away nutrients and plants use them up. **There are two special processes that make growing possible: photosynthesis and the nitrogen cycle.**

Through photosynthesis, plants combine carbon, hydrogen and oxygen to produce food in the form of sugar. This food is like trapped energy from the sun. It is a stored energy that can be used by plants for their growth and reproduction and is also used by microorganisms, animals and humans. One of the primary functions of every plant is to extract carbon from the CO_2 in the air by photosynthesis.

Photosynthesis requires carbon dioxide (CO_2) and the atmosphere does not contain enough, because the air is primarily nitrogen 78% and oxygen 21%.
Supplemental CO_2 must come from the soil.

Photosynthesis takes place within individual chlorophyll units called chloroplasts. These chloroplasts combine the energy from the sun with CO_2 and water to form a six carbon sugar called glucose:

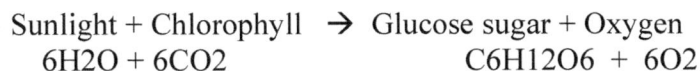

Sunlight + Chlorophyll → Glucose sugar + Oxygen
$6H2O + 6CO2$ $\qquad\qquad$ $C6H12O6 + 6O2$

This process releases oxygen which is why it is good to have living plants in our homes to keep the air we breathe low in CO_2 and high in oxygen.

A chloroplast's chemistry is:

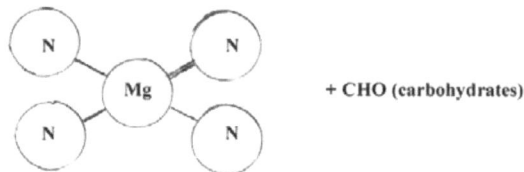

+ CHO (carbohydrates)

The chloroplast is nitrogen intensive and completely dependent on magnesium. However four other elements are also involved. Manganese, iron, sulfur and zinc all play a role in the formation of chlorophyll. Boron is also needed in the movement of sugars within the plant and thus also plays a significant role in photosynthesis. During daylight a plant manufactures sugar and at night a trapdoor opens to permit the chloroplasts to drain their accumulation of sugar for growth and maintenance. Boron is the necessary ingredient to cause the trapdoor to open for sugars to be transferred. If the sugars cannot move out, the plant becomes very inefficient and a backlog of sugar develops.

The second magical process provided by nature is the nitrogen cycle.

Nitrogen is the main element in the atmosphere and is the most abundant nutrient required for plant growth. It is the only plant food derived from the air. It is a vital component of both photosynthesis and the nitrogen cycle and forms 16% of all plant proteins. The management of nitrogen is one of the most important issues to understand for profitable, sustainable farming. Good nitrogen management increases crop quality. Management means getting the right nitrogen balance, as excess nitrogen in the early stages of most crops creates plant top growth out of balance with root growth. Excess N can cause the top to outgrow the ability of weakened roots to supply nutrients. This will produce serious yield-reducing potential, reducing profits. You can also get nitrogen deficiencies showing up as reduced plant size, reduced number of leaves and yellow or reddening of leaves. Nitrogen is also easily leachable from the topsoil.

Whilst nitrogen from the air is free, since World War II nitrogen has been sold to farmers in increasing quantities by the petrochemical industries. Petroleum-based nitrogen fertilisers have increased sales by over 1000%. Commercial nitrogen is made from natural gas which is used to make ammonia which is the primary ingredient in urea and other nitrogen products, so with the increasing prices of oil and gas nitrogen fertilisers are some of the most expensive fertilisers to use. Therefore it is worth understanding the nitrogen process in the soil to minimise your need to buy nitrogen.

The nitrogen cycle is a powerful force that provides nitrogen that is largely free. Whilst nitrogen gas makes up 78% of the atmosphere "plants cannot use pure nitrogen. They use mainly nitrate (NO_3-) and ammonium (NH_4-). Ammonium is a cation (positively charged) and can be held by soil particles but nitrate is an anion (negatively charged) and is subject to leaching.

The nitrogen cycle.

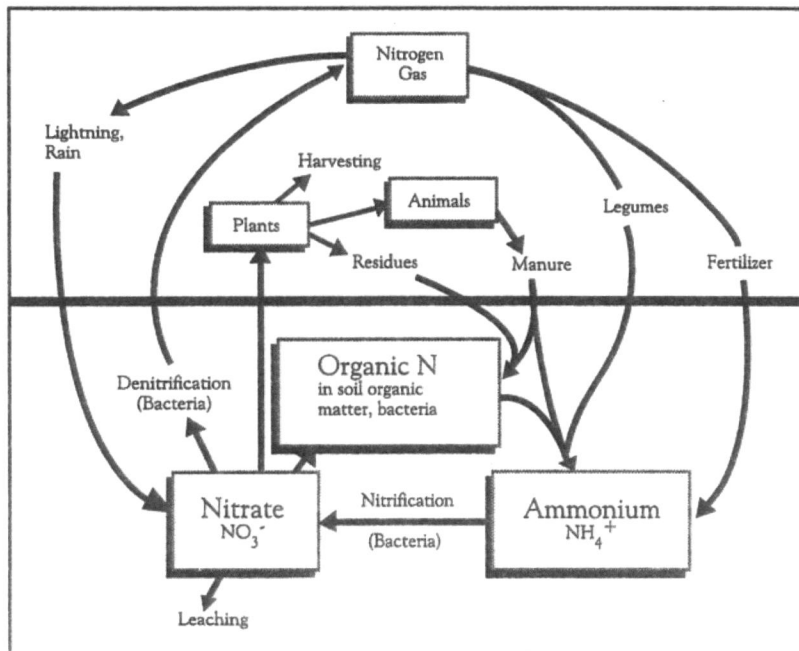

The earth's nitrogen is constantly changing from one form to another and moving from the air to the soil, plants and animals in the above nitrogen cycle." (Diagram and quote from "The Biological Farmer" by Gary Zimmer with permission from the author).

Nitrogen comes from three sources: the atmosphere, organic protein (decomposed by soil microbes) and from fertilisers.

The key step in the N cycle is the uptake of nitrate through the roots and it travels unchanged to the leaves. Nitrate is changed to amine (NH2) so that it can be used for synthesis into amine acids and protein.

Denitrification destroys N and lowers yields. It is carried out by anaerobic bacteria and results in a loss of N from the soil as it is converted to gaseous forms. To avoid this loss, you need to build healthy soils that are highly oxygenated because oxygen kills anaerobes or forces them into dormancy.

Nitrogen fixation is the conversion of nitrogen from the atmosphere into a form available to plants. This fixation process involves an enzyme called nitrogenase which converts atmospheric nitrogen to ammonium nitrogen. This enzyme contains **iron and molybdenum. Cobalt** is also required for nitrogen fixation. A phosphate compound called ATP provides the power source for this enzymatic reaction. The released ammonia is converted rapidly by bacteria to NO3- or is used directly by microorganisms or plants. If anaerobic conditions are created the NO3- is denitrified back to atmospheric nitrogen gas.

Nitrogen management:

1. Ensure your soil has adequate levels of calcium, sulfur and magnesium. **Calcium** is critical for N uptake. **Sulfur** is necessary for the conversion of nitrates into essential amino acids. **Magnesium** levels in the soil also affect N requirements. If Mg is too high, soils require at least 50% more N to do the job.
2. Apply nitrogen-rich fertiliser in the form of animal manure, compost or urea during plant growth when using rows. It is vital to put N on both sides for enhanced nutrient recovery.
3. Applying manure and compost helps increase the soil cation exchange capacity and provide a source of organic matter and trace elements.
4. Plant legumes because the major nitrogen fixers are bacteria called Rhizobium that are found in root nodules of leguminous plants.
5. Combine **humic acid** with all N inputs to stabilise and magnify N and other fertilisers.
6. Remember that if you use **urea** your source of nitrogen, it is expensive. It also causes another problem. Once in contact with the soil it can be converted to ammonia and heavy losses are likely as the ammonia floats away. This loss is minimized if rain occurs soon after application.

The magic formula

When we show visitors/trainees around our farm, the question we get asked the most is "How do you stop the birds and animals eating everything?" We use no netting, electric fences or pesticides. The reality is that birds and animals are not interested in food that is healthy. Their job is to get rid of rubbish i.e. the unhealthy plants. This is Newton's Law of Survival of the Fittest in action.

The magic formula is that we have been taught by Graeme Sait and Bruce Tainio that if you have a healthy balanced soil with a pH of 6.4 then you get no pests or diseases.

We have seen the amazing difference. When we started farming ten years ago we got plenty of pests and diseases. I lost 50 avocado trees in my first year. We had hares and bush turkeys eating our lettuces and carrots. We put an electric fence around our vegetables early on: the hares went under or over with impunity and we were the ones that got the electric shocks.

In the past five years we have had virtually no problems. We have a family of hares live on our property and now they play around here but go and eat next door. Bush turkeys come and have a look and eat only in the compost bins. If we get a problem it is very localised and indicates there is a soil fertility problem in that area and we need to fix it. Our avocado trees were very sick and we were losing a few each year to phytophera as our drainage is not perfect and they grow in a clay based soil. Now they have never looked healthier and our avocado production is rising each year.

6.4 is the magic number.

Chapter 15
Soil building activities: mulching/mounding/composting

"The health of the people is really the foundation upon which all their happiness and all their powers as a state depend."—Benjamin Disraeli.

1. Mulching.

With an organic approach one of the keys to success is the use of mulch.

The topsoil is the environment's recycling factory, alive with bacteria and worms that breakdown and recycle the residue of a previous crop or whatever other organic matter is there. "To plough up the soil is like putting a bulldozer through your factory. If instead you spread mulch on the soil, you are both protecting the factory and supplying it with more raw materials" (See reference 1). The benefits of mulching are:
- o It encourages micro-organism numbers and diversity.
- o It increases carbon content and allows carbon to be recycled.
- o Grass, hay or weed cuttings add fertility to the soil.
- o Mulch reduces evaporation, stores water for the crop to use because it absorbs water. Each organic cell can hold five times its weight in water.
- o Reduces weed growth.
- o It adds nutrients back into the soil.

By comparison, ploughing:

- o Accelerates the rate of water evaporation
- o Increases leaching
- o Encourages weed growth
- o Increases carbon and nitrogen to be turned into gas and released into the atmosphere, particularly when the soil is ploughed after it is wet.

What types of mulch are best?

Some of the best are hay, woodchips, sugar cane and grass/weed cuttings. Woodchips take a relatively long time to breakdown and we use them under fruit trees. With our vegetables we mainly use straw or sugar cane mulch, as it is readily available and not too expensive and comes in easy to move bales. Mowing so that cut grass is constantly piled up under trees increases the organic matter and reduces weed growth.

2. Soil Preparation.

Soil preparation is one of the keys to success for all plants. With trees you should ideally start the process a year before planting to ensure the soil has plant food available when you do plant, as it takes up to a year for elements to be broken down by the soil life. With fruit trees the ideal is to do this soil preparation a year before planting, as this gives time for the soil life to breakdown the elements into plant food and to breakdown the cardboard and paper.

Where we plan to grow crops on land that has not been cropped before or in recent times we carry out a three step process:

a. We put down fertiliser on top of the grass/weeds, usually three or four items, such as chicken manure, rock dust (for trace elements and to help the soil structure), gypsum (to break up the clay) and a general fertiliser such as Nutristore Gold (from NTS which has most of the major elements in the right balance plus more trace elements). You could also use a seaweed or sea salt based fertiliser for all the trace elements.

b. We put down three layers of newspaper and/or cardboard to block out the sunlight. This stops the weeds growing and turns them back into nutrients.

c. Lastly we put on a thick layer of mulch. This has the benefits mentioned above and it stops the cardboard & paper being blown away.
We do not always plan well enough or find enough time to do this a year ahead but doing it prior to planting even when it is done only a few months ahead is better than not doing it at all (see photos 16 &17).

When it comes to plant the seedlings, you do not need to add any more fertiliser. You do need to water when you plant. This is important. Put water in the hole before planting and soon after planting water again. The first seven weeks of establishing a tree or bush are critical to the longterm success, so make sure they get regular watering.

If you have prepared the soil well, you will not need to put fertiliser on the ground for a while. The best way to fertilise new plants is:

a. soak the roots just before planting in a root booster solution made up of fulvic acid, liquid kelp and an all purpose liquid fertiliser. We use Nutri-Tech Triple Ten. This is because the first seven weeks of a tree's life determines the final yield. For vegetables, the critical period is the first seven days when the root system is being established.
b. do a foliar spray every two or three weeks to give the plants nutrients at their time of growth in a way that they can take up the nutrients immediately. This is preferable to putting a fertiliser on the soil because fertiliser takes months to breakdown and become plant food.

3. Mounding.

Mounding your crops helps improve drainage. For trees mounding has several advantages. You need more soil depth for trees than for vegetables and mounding improves the soil depth. It also marks out the rows so that irrigation pipes can be run along the top of the row instead of having to be put underground. This makes irrigation maintenance much easier. Ideally the mounds should run along the contours. We also mound our vegetables to help delineate the different rows of crops and the area between rows provides space to walk without damaging the crops. Mounding reduces water run-off.

4. Composting.

Composting is the process of allowing fresh organic matter to decompose into humus before adding to the soil. It can take from three to twelve months to be formed. Compost can be put out on the soil at any time without harming crops and is like instant plant food. A healthy soil is a living soil rich in microbial activity. To achieve a high density of microbes and bacteria in compost you need to aerate the compost by moving it regularly to add air to the mix.

You can now acquire compost bins that aerate the compost and reduce the period it takes to breakdown the compost to nine to twelve weeks. Sealed bins increase the temperature in the compost as well as protecting against birds and animals foraging. Some have a collection tray that keeps the liquids formed; these are invaluable and when mixed 50/50 with water provide one of the best fertilisers. (Look for the name "Aerobin").

There are several types of compost:
- In bins or drums
- In rows 1 to 2 metres high and 1 to 2 metres wide
- Static or turned periodically to aerate. Ideally it should be turned every month or so.
- In pits in the ground.

You can add manures, grass cuttings and 10% clay. You can add a little more water when turning to help the microbial action.

Benefits of composting

- Increases organic matter in the soil
- Creates a faster rate of nutrient cycling
- Disposes of plant and manure residue
- Produces new compounds that improve plant growth
- Provides food for your microbial workforce
- Improves soil structure

5. Weed control.

The heavy use of mulch and cardboard described in the above section on soil preparation has worked well for us with vegetables, and when planting 100 limes or 200 blueberries. However in bigger areas or with existing farms with more trees or bushes it may not be the most feasible approach. An alternative for these situations is the use of cover (or smother) crops to crowd out the weeds and help fertilise the soil between the existing trees or shrubs. For more information on types of cover crops I recommend you refer to Gary Zimmer's book "The Biological Farmer" (see reference 2).

Chapter 16
Maintenance activities: fertilising/watering/pruning

"The trouble isn't what people don't know, it's what they do know that isn't so."—Will Rogers.

Your biggest asset is the soil. The important part is only a foot deep, because that is where 98% of all soil organisms live and work. After mulching and composting to maintain the level of organic matter, two key jobs to maintain a balanced soil are fertilising and watering. Also in chapter 13 we saw the importance to plants of having an abundance of healthy leaves for maximum photosynthesis, and therefore pruning is an activity to be approached with caution. Pruning can be necessary for longterm productivity of fruit trees.

A. FERTILISING.

The three ways to fertilise:

1. **Applied directly on the ground.** This is the most important and essential method for developing a healthy balanced soil. Rock dust, chicken manure and natural fertilisers in the form pf granules have the best longterm impact in that they can affect a change in the soil balance and pH. It may take up to a year for that impact to take effect, as the fertilisers have to be converted into plant food by the wildlife living in the organic matter in the soil.
2. **Liquid fertiliser on the ground.** This method has a much quicker impact on plants but has less longterm impact. It can be done quickly and efficiently with a fertigation system built into the irrigation system. Rain and irrigation do however leach away nutrients and liquids.
3. **Foliar sprays onto the leaves.** This is the quickest way to affect trees and vegetables. It is almost immediate in its impact and is therefore an excellent way to improve the size and quality of the current crop. You can do a leaf analysis to identify any deficiencies and can correct them straightaway with a foliar spray.

In practice you need to use all three of these methods to get the best results. We use fertigation with liquid fertilisers most frequently.

What do you fertilise with?

Natural animal manures are a good addition. This can be chicken manure, sheep, cow, horse etc. We have used all four and regularly get a truckload of chicken manure. The main thing we have learned is to have a place to unload that is a long way from the house. We have also had trucks bogged twice—with the weight of the manure which is pretty heavy—so choose a spot that is dry or else put in plenty of roadbase to improve the traction for heavy trucks.

The good thing with natural animal manure is it will have organic matter and microbes in it. The bad things are the smell and the fact that you do not know what else is in it, although chicken manure has plenty of nitrogen and ammonia. Caged chickens represent over 90% of all chickens and might have antibiotics, arsenic &/or hormones in their diet. Many of the "baddies" in chicken manure sourced from commercial outlets are removed by bacterial activity when the manure is allowed to compost down.

The basic answer to the question of what to fertilise with is that you need all the natural elements in the right proportions. There are 92 elements that comprise the earth's soil and **the best source is the sea**, as all the nutrients have been washed into the sea over millions of years. Seaweed, fish and sea salts restore the balance and nutrient density better than anything else. They provide good nutrition for plants.

Man-made chemicals are very simple in structure. They act quickly and boost growth. They cause rapid breakdown of humus levels in the soil, releasing the nutrients in the humus and making these available to plants. In the short term this appears to provide benefits in terms of productivity. The long term impact is that it causes rapid depletion of humus levels and of soil micro-organisms. Soil structure disintegrates rapidly.

Volcanic soils have had their minerals replenished from deep down in the earth and we are fortunate that our farm is on Tamborine Mountain, which is a volcanic outcrop. Over the long term soils that have been leached of their nutrients are renewed by Nature not only by volcanic eruptions, but also by advances and retreats of glaciers and by slow uplift of the earth's crust, such as in the Great Dividing Range in the east of Australia and areas of South Australia. As much of Australia is very old, most of Australia's soils are poor in nutrients. The good news is that soils that lack minerals can still be made productive. A brilliant example of this is the south-west corner of Western Australia where the soils are very infertile, yet this is one of the agriculturally most productive regions of Australia with hundreds of vineyards and a massive wheat belt. Farmers in this region have to bring in huge quantities of fertilisers.

Plants remove nutrients from the soil. Irrigation and rain leach nutrients out of the soil. This is why you need to regularly fertilise to put back the nutrients. Different plants take out more of some nutrients than others, so the only way to know with certainty what to fertilise with and in what quantities (to maintain the ideal soil balance) is to do regular soil tests. It is strongly recommended that you invest in at least an annual soil test. Once you know what to put on, you either put the granules on the soil (by hand or with a spreader) or get soluble fertilisers and put them into a fertigation system.

You need to send your soil samples to a testing laboratory to do the actual soil analysis. One of the reasons for doing soil tests underline annually underline is that it takes a year to change the pH and soil balance. It takes time for all the microbes and worms to break down the granules and convert them into plant food.

Because of this long timeframe, a major factor in the success of your crop growing is the use of foliar sprays. By spraying liquids onto the leaves of plants and trees, you can have an almost immediate impact. This is highly desirable to ensure you get good productivity. How do you know what is needed in the sprays? You do a leaf test. This can be done with both vegetables and fruit trees. I use the same laboratory at Lismore to send the leaves to. You collect the leaf samples and send them off early in the week (so that they are not sitting around over a weekend) to arrive fresh. The laboratory sends a copy of the resulting analysis to NTS and they prepare a report with useful comments and recommendations.

You then get the liquid elements needed and make up a spray solution with water plus an ingredient called "Cloak Spray Oil" which helps the ingredients to stay on the leaves.

The five magic ingredients for nutrient-rich crops.

1.All living processes are reliant on a source of energy. That energy is **a**denosine **t**ri-**p**hosphate or ATP and **phosphorus (P)** is the key element in ATP. Good P levels in the soil mean good energy levels which provide the fuel for growth and production. The most vital time in a plant's life that determines final yield is the stage of early root development. P is essential for vigorous root growth. Another peak energy time is when plants form seeds. You need a soluble phosphate to boost root growth at the time of planting.

Phosphorus does not leach so it is often the case that there can be an excess of P in the soil. This excess comes partly from it being one of the three widely used chemical fertiliser ingredients in NPK (the other two being nitrogen and potassium). But P can get tied up in the soil and not always be readily available to plants, as in its simplified chemical fertilizer form P quickly bonds with other elements in the soil.

You need a balance of other minerals such as magnesium for good P uptake. One of the main reasons P gets locked up is the poor pH of many soils. pH plays an important role in P availability with P most available between pH of 6 and 7.

The CSIRO estimated that there is approximately 10 billion dollars worth of P locked up and unavailable in Australian soils. The key to unlocking this potential is the use of organic matter and phosphorus-solubilising microbes. **Humic acid** stabilises P so that it does not tie up with other elements.

Liebig's Law of the Minimum states that the nutrients in least supply govern plant growth and in Australian soils this minimum or limiting factor is P. It may be present, but it is not available to plants.

Routine leaf analysis is the best way to identify whether P is getting taken up from the soil. Foliar sprays help correct the problem quickly and keep productivity up, but the longer term solution is to build organic matter and microorganisms.

In your soil test you only need 50 to 70 ppm of P and your test will often show much higher levels but this does not mean that the P is available for the plants. In 2005 our soil test on our vegetables showed P at 540 ppm but in the NTS analysis showed that the immediately available plant food was low and microbial activity, particularly fungi activity, showed a very low reading. It is fungi that are largely responsible for the stabilisation of both calcium and phosphate.

2. Silicon is a key element and is present in the soil combined with oxygen as silica. A third of the earth's crust is made up of silica and it is the second most abundant element on the planet but it is often not in a plant available form. It may be abundant generally particularly on our beaches but still be at low levels in our farmland. Furthermore most humans have a serious silica deficiency, because of this lack of uptake by plants. Rice accumulates up to 16% silica (a reason why the Japanese are a healthy nation with long lives). The ideal level of silica or silicon in the soil is only 100 parts per million (ppm). My early soil tests all had levels around only 40 to 60ppm. Silica is important because it increases photosynthesis, strengthens cells thereby increasing their resistance to pests and diseases and it improves yield and the quality of produce by helping in the uptake of nutrients.

Silica has a major impact on photosynthesis, because the structure of the leaf and the structures on the leaf are both silica dependent. The leaf is like a solar panel capturing both water and sunlight on the topside and CO_2 via the stomata openings on the underside. The better the leaf presents itself, the better the photosynthetic potential. By strengthening the plant it creates greater tolerance to environmental stresses. It also increases root health and development. Yet another benefit that comes from silica is that it helps release phosphorus and other nutrients that can get locked up in the soil. So you need a source of silica for your soil and crushed lava or cracker dust is one. For rapid release a liquid fertiliser called "diatomaceous earth" can be used (see sources in Appendix C).

3. We need help to manage effective mineral uptake and one key material that does this is **Volcanic Zeolite,** a special type of clay with a very high nutrient storage capacity; it also has far greater strength and stability than clay. Zeolite crystals have a three dimensional structure like a honeycomb with interconnecting tunnels and cages. These can store up to 70% of their dry weight in water, 30% in gases and up to 90% of certain hydrocarbons. Zeolite has a CEC of 140; it can function as a water and nutrient reservoir in the soil indefinitely. Zeolites have a particular affinity for ammonium nitrogen and potassium. These two elements are often poorly stored in many soils so this is a big plus. This is especially important in the case of nitrogen. A healthy well balanced soil contains 75% ammonium nitrogen and 25% nitrate nitrogen; however in most soils this ratio is inverted in favour of nitrates. Nitrates are proven carcinogens for animals and humans and in plants are a calling card for insects. Improving nitrogen management is a major priority for all food producers as we saw in the last chapter.

Granular and liquid forms of zeolite are commercially available.

4. Kelp or seaweed is the best source for all the trace elements. Seaweed powders can be purchased for use in fertigation systems or foliar sprays. To correct most mineral deficiencies a foliar spray with a seaweed fertiliser is like a magic recipe to improve plant growth and yields.

5. Rock dust is another great addition to the soil to help make sure there are plenty of trace elements present. It also improves the soil structure and makes the soil easier to work. Some sources are covered in Appendix C. Granulated basalt or crushed lava are good. We saw the amazing impact of rock dust on one area growing rhubarb that had no rock dust. An adjacent area with rock dust had rhubarb almost twice as tall, twice as thick and a darker red.

What is fertigation and how do you do it?

Fertigation is one of the best time saving investments, unless you have a very small area to fertilise. Putting out fertilisers by hand or even with a spreader can take a lot of time and it is difficult to ensure an even spread. Assuming you have an irrigation system for watering, a fertigation unit (see photo 37) is installed between the water source and the sprinkler system. The unit is basically a drum or tub that is inserted into the irrigation system that enables you to input liquid fertilisers into the water flow and get them spread onto the ground through the sprinklers. Being liquid rather than granules it has a quicker impact and has the benefits of ensuring the same amount goes onto each tree and saves a huge amount of time.

We put one in for our avocados as we have 250 trees to look after and the cost was about $2000. We installed the avocado fertigation unit in 2005. The value of our avocado production jumped in 2006 by 66%. It has been so effective that we soon decided to put another one in for the vegetable area, so we now have two fertigation units. To make sure it is effective you need to clean out both the fertigation system and the sprinklers for blockages periodically

What fertiliser to put on is determined by your annual soil test and regular leaf tests. When you do a leaf analysis the recommendations from NTS usually give you not only the inputs for foliar sprays but also supplements recommended for fertigation. Whilst you fertilise more frequently the cost is offset by the labour savings and increased production and revenue.

The next question is how often should you fertilise. If you are serious you need to do it frequently. Plants like humans need regular nourishment, so we try to fertilise monthly. I suspect that every 2 to 3 weeks would be even better but there is always more to do than hours in the day, so once a month is a realistically achievable goal.

With fertigation you need to periodically check your sprinklers to ensure they do not become blocked. Some fertilisers are not as soluble as you would like and you should stir the fertigation mix and put the fertiliser into the unit through a sieve. My fertigation unit needs to be cleaned out regularly. Your local pumps and irrigation shop should be able to supply and install one.

Foliar sprays.

This is the way to ensure nutrient-rich crops and good productivity. Ideally at least four times a year do foliar sprays of both fruit and vegetables. It has been proven conclusively that nutrients get taken up through the stomata of leaves via foliar sprays. These can provide a significant increase in yields because they provide new supplies of essential nutrients at the critical time of need. Seaweed or fish-based foliar sprays make sure all the trace elements are available. Foliar sprays are more efficient than soil applied fertilisers because they don't suffer from leaching and have a quicker impact on the plant.

I started out with a backpack with 100 litres capacity and water is very heavy and 100 litres is not enough to do 250 trees so you need several trips and it is very tiring particularly on your arms. I then moved up to a spray unit that fitted in the trailer behind our rideon and runs off the rideon battery. This was much better but I still needed to stop the rideon twice for each tree and walk around the tree spraying it.

My next phase was to get a bigger more powerful spray unit that works off the PTO unit on our tractor and is fitted to the three-point linkage at the back of the tractor. This was much more effective and it was possible because of the extra power and farther reach of the spray to do the trees whilst driving along. I would drive down a row between the trees and do one side and then come back up the next row and do the other side. It was possible to do the 250 trees in 2 hours with this approach.

I found two even better solutions. Next I got someone to walk beside the tractor and handle the spraygun while I drove the tractor slowly. I like this approach and I also borrowed the spray unit so I did not have to buy one. A second solution that is also pretty effective is to contract someone to do it for you. This is much less tiring.

Before doing a foliar spray, I get a leaf test done so that I know what nutrients the plants need now. Sometimes a soil test will show that specific elements are present in the soil but the leaf test shows that they are still deficient in the leaves and fruit. This is because some elements get "locked up" in the soil (such as phosphorus) because either the balance is wrong or there are insufficient, or wrong type of, microbes/worms/fungi to convert them to plant food.

With the mix to be sprayed on the leaves I generally add some "Cloak Oil" and some fulvic acid to make it more effective. Each year at the end of winter/start of spring I do a special boron foliar spray particularly for the fruit trees. This is done before flowering to increase the number of flowers and help the fruit set. Graeme Sait says "there is no crop that will not benefit from a boron foliar spray before flowering and use some fulvic acid in the spray mix."

Foliar sprays are best done early in the morning when, I am told, birdsong opens up the stomata to seven times greater than their normal size. The leaves will take in more nutrients.

B. WATERING.

One of my few good decisions early on was to put in a computer controlled irrigation system for the avocados. This means that trees are watered even when I am away and each tree gets the exact amount you decide on.

There is however much more to learn. The key questions which I am still researching the best answers to are:
- How often do you water?
- When it rains, you turn off the irrigation system. When do you turn it back on again?
- When you get drying westerly winds, do you increase the amount of watering and if so, by how much?
- Should you water more in a normal summer and less in winter? (In the summer of 2007/2008 we had three months of heavy rain in Nov. /Feb.).

I don't claim to have all the answers to these questions.

I started out on the assumption that trees needed some water every day. After further research I found that some authorities recommend watering weekly e.g. see the excellent reference manual called the "Avocado Grower's Handbook." (reference 1). Some recommendations are for twice a week. In our own "Avocado Study" of eight orchards on Tamborine Mountain we found an enormous variation in not only the frequency but also the amount of water put on. Some have no irrigation and rely on nature, but the interesting finding was that the two most productive avocado farmers watered the most. The avocado farmer I respect the most waters generally weekly, i.e. gives a heavy watering once a week rather than less each day.

Whether you should water daily or weekly is related to the soil type and water holding capacity. On Tamborine Mountain although the base is volcanic, there is quite a lot of clay and the soil has a good water holding capacity and does not require daily watering. Sandy soils in hotter regions have a low water holding capacity and could be watered daily. Daily watering on TM exacerbates the risk of phytophera disease. So after 10 years of using a daily cycle I decided to reprogram my computer to water weekly.

There are some more basic irrigation issues to deal with before worrying too much about the above questions. **Firstly**, maintenance of your irrigation system can be a big job. Sprinklers get blocked (by fertiliser, mulch and ants) and leaks occur. I put pipes underground initially because this means you can mow the grass more easily. However it also means you can accidentally cut through a pipe when you forget exactly where they are. It also means that when you have a leak it is harder to find the source and more work to fix it. This is even more important in the vegetable areas where you use a pitchfork to loosen up the soil or to dig up root vegetables—it is very easy to put the fork through a buried pipe. Finally when you want to extend or add to the irrigation system you have to find it first. I now much prefer to run the pipes above ground which saves a lot of time and this is a good reason to plant trees and vegetables in rows, so that you can run the pipes along the rows.

Secondly, when I started, I put in one sprinkler per tree (see photo 27). This is fine for small seedlings and for trees up to say two years of age, but when the tree gets bigger you will find that only one side of the tree gets water. So for big trees you need to put in two sprinklers. Alternatively if you put trees in mounded rows another advantage is that you can have one sprinkler half way between each pair of trees (see photo 20). This sprinkler then can water one side of each of two trees (the trees at the ends of rows still need an extra one).

Another issue is how high should the sprinkler be? I used small sprinklers initially (one foot in height or 30 cm). These are more likely to get blocked or buried by fertiliser and mulch. I decided that sprinklers one metre high are more effective and require less maintenance. They are more effective because as the tree gets bigger it needs water over a larger area. It is not a trivial job to change 250 small sprinklers on underground pipes to 500 taller sprinklers for my 250 avocado trees.

Water monitoring systems

1. The most common is a tensiometer, which is a probe you insert into the soil and it has a gauge which gives you a reading on the moisture level in the soil. You can use two tensiometers at different depths to get a clearer idea of the soil's water content. They cost about $120 in the year 2000.

2. Another newer device is called a "Full Stop" and was designed by Australia's CSIRO and helps you to "see" what is happening down in the root zone when you irrigate. It is available from a firm in Adelaide called Contact Measurement Australia Engineering (MEA) (tel. 08-8332 9044, websites: www.fullstop.com.au and www.mea.com.au).

There are more sophisticated devices which are of course more expensive. You can have the readings fed automatically into a computer that decides for you when to turn on and off the irrigation. This clearly has an advantage if you are away on holiday or not able to check regularly yourself for some reason, but generally these would not be justified unless you have a really large number of trees.

3. There are four water monitoring systems that measure water content and provide input to a control system (usually a computer). Two are portable, the Gopher or the Diviner, and two are non-portable called EnviroSCAN or C-Probe. With the portable versions soil moisture readings can be measured onsite or downloaded into a computer and calculated later. These estimate when to water and how much to apply.

The two non-portable devices are continuous moisture monitoring devices. At set times the measurements from the sensors are relayed along a cable to a data logger for recording. They are then downloaded from the data logger to a computer to provide recommendations for watering.

I started out using tensiometers and these are fine once you understand the significance of getting watering right.

4. There is also a simple "pH moisture meter" that you insert in the soil and get an immediate reading. This only tests a few inches or cms. into the topsoil but has the big advantage of being portable and therefore you can readily test several areas.
I have recently acquired a pH soil moisture meter and will be trying it out. Besides using aboveground sprinklers there are some other options for irrigating. I use a dripfeed irrigation pipe for watering our macadamias and blueberries. Drip feed is also adequate for young trees. It has the advantage that it is cheaper and it is both easier to install and to maintain. You just need to put the bushes or trees in rows and run the drip feed pipe along the ground along the rows. Drip feed provides significantly less water per hour than sprinklers, which generally do between 55 and 65 litres per hour. Drip feed provides only two litres per hour.

Sprinklers can vary in the diameter of the circle they cover, so as the trees get bigger you can increase the area covered by making a change to the sprinkler head.

There is yet another option for irrigation. Not only can the pipes be under or above ground (when the sprinklers are above ground in both) but the pipes and the watering can both be underground. This radical idea is proposed by Peter Andrews in his wonderful book "Back from the Brink" (reference 2). He claims that as much as 90% of irrigation water can be lost due to evaporation leaving salt behind. He contends that irrigation water evaporates at ten times the rate of rain water.

To irrigate sustainably you need to minimise evaporation. Peter Andrews proposes a series of leaky pipes laid at a certain depth that maintains a water table at that depth rather like a drip feed system laid underground. When the irrigation is turned on the water goes into the ground to wet the roots of the plants using only about 15% of the water that normal irrigation uses. This proposal makes a lot of sense for the hotter areas of Australia.

C. PRUNING METHODS.

There are two basic types of trees: deciduous and evergreens, and pruning methods vary for each. Avocados, Citrus and Macadamias are all evergreens. Blueberries, Apples, Pears and most stone fruit are deciduous. Pruning's main aims are to control the height and shape and to let more light in to help increase flowering.

a. Deciduous pruning.

The purpose is normally to shape the trees and there are two main shapes advocated:
the vase and the pyramid.

1.The vase is the older more traditional form, but produces a weaker tree and branches often break under the weight of their crop. The vase takes up more room and is less productive than the pyramid.
The centre is removed to allow maximum exposure to sunlight.

2. The pyramid is a more stable but taller format, focussing on a strong central leader. Three or four main branches are selected and all extraneous growth is removed. The lower selected branches are kept longer than the upper ones.

Most pruning is done in winter with only a general tidying up in summer.

Six guidelines for pruning:
1. remove branches that do not conform to the desired shape of the tree, especially over-vigorous upright shoots and all inwardly directed growth;
2. remove weak or old shoots;
3. remove any dead or diseased wood;
4. thin out old fruiting spurs;
5. thin out crowded growth so that all shoots remain well spaced;
6. shorten back last season's growth by two-thirds to encourage spurs to form and to keep these spurs close to the main branch.
(See reference 3).

b. Evergreen pruning.

I will illustrate what I have learned from avocados over the past ten years. Avocado tree growth is rapid and trees commonly reach 20m in height. It becomes uneconomic to harvest and spray large trees. Fruiting takes place on the perimeter, and as trees crowd each other, light penetration into the orchard is reduced and the fruiting surface migrates to the tops of trees. This reduces fruit size, yield and quality. When we prune we have learned that if you leave the cut branch exposed, disease can enter; to stop this happening you paint the cut with white paint.

Avocados have a long cropping cycle. Fruit can be carried on the tree for over 12 months. Trees often carry an old and a new crop at the same time. This means there are few opportunities for pruning that do not risk damage to fruit and yield.

In our own four year avocado study of eight avocado orchards on Tamborine Mountain we found **two** sensible methods of pruning. Our best avocado grower does some pruning every year, usually straight after picking (often in October). He prunes fairly severely, usually about one fifth to one third of each tree. His trees are fully mature (over twenty years old) and quite tall, requiring a cherry picker to harvest. The benefit of this method is that his production is not adversely affected by pruning .

The second method is the one I use, which we learned at a Queensland DPI meeting we attended early in our experiences with avocados. A South African researcher, called Clive Kaiser, on a visit to Australia gave a talk which emphasised the huge loss of fruit and flowering when trees overcrowd, as all trees with the exception of those on the outside will flower and set fruit only on the tops of trees. This highlights why you have to prune if you want to maximize fruit numbers. Trees need to be exposed to light on all sides and when they are they will flower and fruit on five sides (i.e. the four quadrants and the tops of trees). When overcrowded trees are left unchecked, they simply grow vertically and set fruit only on the tops which make the fruit unattainable as avocados can grow to amazing heights.

This is why many large avocado growers use cherry pickers to reach the high fruit. If left unpruned avocado trees can grow too high for even cherry pickers to be able to reach the fruit. When the fruit get too high some growers remove some trees totally to bring in more light to the other trees or else cut trees right back to a stump (called staghorning) or do severe pruning.

 We decided we would not use cherry pickers or ladders to pick our fruit (expensive, dangerous and time consuming) so we adopted Clive's recommendations. We keep the trees pruned to about seven feet or just over 2 metres in height so that we can readily foliar spray and pick from the whole tree without ladders or cherry pickers.

Avocados are an evergreen subtropical fruit tree. Most pruning information is based on commercial deciduous fruit tree research and these are very different, particularly where flowering is concerned. Most deciduous fruit trees produce flowers along the branches providing light is able to penetrate the canopy, so their pruning allows light penetration into the canopy e.g. cutting the central leader or making an open vase effect. Avocados flower on the periphery, so this (deciduous tree) approach does not work for avocados. Contrary to popular belief pruning an avocado to allow light into the centre of the canopy does not result in more flowers but produces vigorous vegetative growth. This loses production over two seasons. This "window pruning" will only result in flowering on the extremities of the regrowth in the following year.

The key principle Clive advocates is based on the fact that the more terminal buds there are, the more potential fruiting surfaces there are, so the **concept of increased tree complexity** is of paramount importance to increased fruiting potential. This principle applies to all fruit trees. Buds must be exposed to light from mid-summer (or end-February in S.E.Qld.) for flower induction to take place. Regrowth needs to be encouraged to branch as much as possible from as low as possible by selective pruning cuts and by tipping the shoots. The timing is important. Regrowth should be controlled on a regular basis during the spring and summer months (the two annual flushes). Pruning and tipping should cease in late February as flower initiation begins shortly afterwards. By removing the tips of growth (the apical bud), the two and sometimes three buds, which remain are forced to grow, thus increasing tree complexity.

Chapter 17
Seven Free Workforces.

Q. How many people work in the Vatican?
A. About half.
Answer attributed to Pope John XXIII

In our first ten years of farming we survived and grew our business with no paid employees. We obtained our labour "free" from seven sources:

1. We embraced the Wwoof programme—**W**illing **W**orkers **O**n **O**rganic **F**arms. This is an international organisation used mostly by university students and backpackers that enables them to obtain free food and accommodation in return for four hours work per day. There is a wwoof website www.wwoof.com.au where hosts can also put up their details with photos. It says, "Want to arrange a farm stay in Australia for work instead of money?"

We have had over 200 woofers stay with us over the past four years. Some have become good friends, most stay for two or three weeks, and a few have stayed for months. We have made good friends from all around the world (see photos 35,38 &41).

Their expertise varies greatly but with supervision you can get a lot done with three or four willing workers. It costs $50 per annum to be a registered host and to have your contact details and a description of your property/activities published in the Australian yearbook. Generally the woofers telephone and you have a chance to vet them, so you need to have a chat and check out their experience and most importantly get to know whether their English is adequate. It is very difficult to get any work done if there is a language barrier.

One of the aims of the wwoof organisation is to teach people about organic growing, bio-dynamics or permaculture, so they do not encourage conventional farms and learning is mainly by doing and the wwoofers asking questions.

It is a great scheme and has been and still is a boon to us in providing helpers. When it rains we get woofers to help make jams and chutneys or do packing or cleaning.

2. The next free workforce is invisible. It works in the topsoil and is best described by Graham Harvey (reference 1): "Just a teaspoon of healthy soil contains over 5 billion living organisms, representing 10,000 or so different species" of bacteria, microbes, fungi and worms.

Not all soils are healthy. The microorganism workforce is only available to healthy soils. Conventional or industrialised farming uses huge quantities of pesticides, herbicides and fungicides which kill off this workforce. If you use compost and mulch and animal manures, you will provide a home and food for millions of micro-organisms who will convert the elements into food for your plants and aerate the soil.

This workforce makes all the difference between producing poor food lacking nutrition, but with plenty of toxic chemicals, and good nutrient-rich food. **Microbes work 24 hours a day for you for free.**

There is an important piece of knowledge to maximise the effectiveness of this workforce. A key ratio in healthy soils is the bacteria: fungi ratio and this varies depending on your crop. From the Nutritech Certificate of Sustainable Agriculture course I learnt that all soils vary and have different ratios depending on the different plant species growing in that soil. You benefit from diversity; it is particularly important to encourage beneficial microbes and fungi. This is why providing organic matter rich in compost, humic and fulvic acid etc. is critical and conversely why pesticides, fungicides and herbicides are so negative with regard to growing nutritious food. Vegetables tend to want fungi: bacteria ratios around .5 to 1:1 whereas trees, bushes and strawberries favour ratios from 5 to 100:1.

The reason why strawberries prefer fungi dominated soils is that they occur in nature on forest floors where the soil is fungi dominated.

To adjust this ratio you can do two things. First, you need to provide the right type of food. To increase bacteria you provide more green material, fulvic acid, molasses and nitrogen. To increase fungi provide straw, sawdust, bark, humic acid and kelp. Secondly, you can add the ones that are missing with a special microbe or fungi brew. This process is called inoculation.

3. The third workforce also lies beneath the ground. This is the root structure and web of root hairs that supports and anchors your plants. These seek out the nutrients that plants need from the soil and help aerate the soil.

4. The fourth workforce is visible. These are the leaves of plants and trees that like the roots interact with the environment. They are the solar panels that grab the energy from the sun and initiate the photosynthesis process.

5, 6 & 7. Air, sunlight and water are three workers. They do not get paid. They do not need man's help and existed well before mankind had its love affair with man-made chemical fertilisers, pesticides and herbicides. With their help plants and trees flourished without man's "creative" money-making inputs. Together with the invisible microbes and bacteria they create bountiful nutrient-rich quality fruit and vegetables.

Their roles and the work they do come via nature's two magical processes—photosynthesis and the nitrogen cycle.

Chapter 18
Avocado Study

"Insects and diseases are the symptoms of a failing crop, not the cause of it," Professor William Albrecht.

This chapter is just about avocados but it illustrates how much there is to learn about each crop and maybe a way to learn more about growing crops in your own area. It illustrates that when we come to subjects like seeds, plants, trees and nutrition there is a lifetime of study involved. Farming is a challenging activity, not only physically but also mentally.

In 1998 I planted 300 avocado trees based on a suggestion from my golfing partner. I had no knowledge of farming or avocados and thought they would look after themselves. They are however not an indigenous tree used to the Australian climate they come from the Caribbean. When I lost 50 in the first year, that triggered me into doing some research to learn more about growing things.

I joined the local producers association and as there are 300 avocado orchards on Tamborine Mountain avocados is one of their main interests in coming together. I went to their meetings where they had regular guest speakers and I was fortunate to hear Graeme Sait speak at one of their meetings, which eventually led to my doing his "Certificate of Sustainable Agriculture" course.

By then I had started to realise how much I didn't know and as avocados were our biggest source of income, I initiated a four year research study and persuaded seven other farmers to participate. We have shared all our information on production, costs, fertilisers, watering, mulching, pruning and so on and have produced a report each year on the results. We sell this report to other members of the Tamborine Mountain Local Producers Association and we regularly get calls from people who have just bought a property asking how to look after their trees.

This study has been a real eye opener on just how much there is to learn on just one crop. We now have benchmarks on what yield to expect, how much fertiliser to put on, how much water to put on, when to prune and how to prune and why you need to prune.

I am still amazed at how much there is to learn just about watering and irrigation. The two most successful farmers in our eight, water the most. The most successful puts on more fertiliser than anyone else. As a result of this exercise I now <u>know</u> that my initial motivating goal of achieving an average of 300 avocados per tree per year is highly achievable—but only if you have the "how to" knowledge and if you apply that knowledge. I should have asked Alec how to look after the trees. Five years of neglect in the critical early growth period of the trees has meant I have to undo those five years of harm. This is difficult to do and takes a while.

However avocados should yield an average of 300 avocados per tree per annum and can average over 400. I now know this is true because I have seen others do it.

One of the key lessons from our avocado study that applies to all fruit trees is with regard to watering. This is the need to move from one sprinkler per tree when the tree is say up to three years of age to two sprinklers per tree, one on each side as the tree gets bigger. (If you plant the trees in rows you may be able to get away with one bigger sprinkler between two trees—see photos). The amount of water required for a mature tree is also much greater than for young trees. I started out doing 15 minutes per day per tree of watering but a standard sprinkler does 65 litres per hour and the real requirement for mature trees is about 1200 litres per week or 171 litres per day which takes nearly 3 hours per day of watering per tree.

In practice there are many variables such as the impact of rainfall, the size of the tree, winds and the seasons. Trees need most water in summer and least in winter. They need almost as much as summer in spring when they have enormous activity (flowering, fruiting and the spring vegetative growth or flush) and autumn needs more than winter.

Ken Pegg from the Queensland Dept. of Primary Industries gave me these rough guides in terms of litres per week.

Diameter watered per tree	Spring	Summer	Autumn	Winter
4m	400L	500L	250L	140L
5m	650L	700L	400L	220L
6m	950L	1000L	550L	300L

Young trees e.g.2 years old require:

	Spring	Summer	Autumn	Winter
	80L	90L	50L	30L

Another factor here on Tamborine Mountain is the impact of westerly winds from the interior. The hot dry westerlies in the summer raise the need for water significantly. In winter we occasionally get bitterly cold westerlies and these also seem to generate a need for more water.

After six years of growing avocados the study made me aware that there is an excellent reference manual on avocados that every serious avocado grower should have. This is the Queensland Dept. of Primary Industries' "Avocado Growers Handbook". If I had studied this before planting any seedlings, my production would have been much higher. Same advice to all growers: research the information available from relevant industry groups before you make important decisions to avoid making costly mistakes.

We pay an annual avocado grower's levy to the Dept. of Agriculture and most of this is used for research. One of the things I would disagree with regarding industry and Agriculture Dept. recommendations is that most of the research is on the management or control of pests and diseases. I think this is unnecessary and wasted expenditure because the solution seems quite clear. If you have a healthy balanced soil with a pH of 6.4 you will not get pests or diseases attacking your crops.

Of course conventional farming uses man-made chemicals and gets sold heaps of pesticides and herbicides, so it is unlikely that their soil will be 6.4 pH. The amount of pesticides and herbicides sold each year increases and the incidence of pests and diseases also increases each year. It seems to me that there is something wrong with the logic. Possibly it's a case of "if what you're doing doesn't work, do it more often", which is I believe a strategy used frequently by those who are mentally ill. Chemical companies fund research into pest management programmes which at least creates some work for the researchers.

The main disease that attacks avocados is phytophera or root rot. Graeme Sait says "No avocado tree growing on soil with a pH of 6.4 has phytophera".

> For the record, as most avocados are not grown in soils with pH of 6.4, the recommendations to minimise the impact of phytophera are:
> - Plant seedlings in soil prepared with chicken manure and gypsum (ammonia helps kill the fungus),
> - Once the tree is mature you can inject the trunk twice a year(after the spring and summer flushes) with phosphonic acid,
> - There are two alternatives processes to injection which have the benefit of being able to be used on young trees (which are too small to inject). These two alternatives are (i) to do a foliar spray with phosphonic acid and (ii) to paint the bark of the tree's trunk with a paintbrush dipped in a phosphonic acid solution that has a 2% penetrant additive (to help the solution penetrate the bark). Two penetrants available are called "Pulse" or "Prosil". You paint from the ground level up to the first branch.
> - If your tree is showing signs of phytophera (such as yellow leaves) you should cut down on the amount of water given.

Research on painting the bark is quite recent but it appears to be more effective than injecting or spraying.

Apart from watering some of the key points learned from this study that apply to all crops are:

Fertilising: you need to reinvest some of your revenue earned back into fertilising. The rewards will far outweigh the cost; a good yardstick is to spend 10% on fertilisers.

Pruning: careful pruning can maintain or improve yields.

Mulching: add mulch to trees every year.

Fungi: trees like lots of (beneficial) fungi and injecting fungi spores can provide a significant boost to health and productivity.

Communication: when eight people do their own thing, they will develop eight different strategies. When you share what you know in a group, everyone benefits. You can choose to incorporate the best strategies. Even the best performer in the group has benefited by sharing and his productivity continues to increase.

Chapter 19
Planning

"Nothing can resist the human will that will stake even its existence on its stated purpose."- Benjamin Disraeli.

To achieve the aim of this project i.e. to create a profitable small farm organically, you need a business plan to clarify your goals and direct how you will achieve them. You need to decide which crops you will grow and where you will grow them, how much land you will need to obtain your targets for each crop. You need to order the seeds and seedlings, the fertiliser, and do the soil tests etc. So you need to have a vision for your property first.

Based on your vision for the property, you need to make some annual revenue projections and costings i.e. a budget, and decide what equipment to invest in and how you will finance these investments. This chapter will give you some guidelines and then chapter 23 provides some of my actual revenue projections as examples and chapter 24 will show you what actual investments I have made in machinery, tools etc. Chapter 25 sets out what steps we actually took and when over the past 10 years or so and some plans we have yet to achieve.

A business plan provides you and your bankers or financiers with confidence (if you need to borrow some money). It should be a living document that changes over time as you refine your goals, and it sets some benchmarks so that you know how you are going in achieving your desired end-result. All plans can be improved, so it is worth sharing yours with one or two people such as your partner or accountant or mentor to get some feedback.

There is no doubt that written goals improve your chance of success. Once you have clear goals with target completion dates, you need to focus on what are the actions you need to take to achieve those goals.

Your plan needs several different timeframes with some goals for say 3 months time, one year, five years and maybe even ten years. I got our accountant to do a ten year cashflow projection once we had a couple of years' actual data and this impressed our bank. We also considered a joint venture and the ten year projection was again a useful tool and aid in the discussion.

Business plans vary in format, content and size but you will need several pages to cover all the key points. Things you should consider when developing a plan for your property include:

- Your vision for your property
- Your key goals
- Marketing
- Action plan of how you will achieve your goals

- Finance/Cashflow
- Succession plan or exit strategy

Some of the more detailed topics to consider are:

- Operations/Legal/Tax structure
- Suppliers
- Finance/ Profit & loss projections/Cashflow/Capital investments
- Staffing & resources
- Competition & Pricing

One of your key goals is to continually improve your soil with ongoing plans to bring in mulch, create compost, do soil tests and use at least once, a prescription blend fertiliser to bring your soil close to the ideal pH of 6.4. This is one of the key principles of success (see chapter 3). The healthier your soil, the fewer problems with weeds, pests and diseases. You need to monitor your soil pH and do regular soil tests. It is a good idea to set a month that you know you will do a soil test every year.

Some of the important goals are your plans for fruit trees such as:
1. Decide which fruit trees and how many of each and where you will plant them.
2. Put in an irrigation system.
3. Prepare the soil.
4. Order the trees.

You will probably need to devote some effort and maybe some funds to marketing your produce. Marketing is normally the crucial part of a business plan and is covered in detail in the chapter 22. However in this case marketing will fall into place relatively easily if you focus on getting a healthy balanced soil, because then you will have such high quality produce that customers will beat a path to your door.

Most plans have three parts. Firstly, **a strategic plan** for the vision you have for your property and your major strategies. The next chapter discusses this. Secondly, you need **an action plan** with the steps you will take to achieve your goals with dates for completion. Chapter 21 provides an outline. Thirdly, **a marketing plan**, which is the subject of chapter 22.

If you need help in this area I recommend you contact Jane Pollard on Mt.Tamborine on tel. 07-5545 2588. She can mentor/guide you through the business plan documentation plus provide ongoing monitoring as your business grows with financial statements, budgets and tax advice if need be.

Chapter 20
Strategic Plan

"Not every end is a goal. The end of a melody is not a goal; but nonetheless, if the melody had not reached its end it would not have reached its goal either."—Nietzsche.

You need to write down your vision for your property, the strategies you will adopt and the major goals with a timeline. I will give you mine as an example. Yours will be different depending on the size of the property and your choices of crops etc. We are all different and while the book has a financial goal as a unifying theme not everyone will want a financial goal. I have several key goals and the financial one is not the major one.

My wife and I started out with two goals for our property, one of being self sufficient and producing nutritious food without using herbicides and pesticides and the other to do training. After a while food became the primary goal for some years. When asked what we do, we developed a three sentence reply:

> *In essence, our business is food.*
> *We grow food, market food, buy food, teach how to grow food & write about food.*
> *We love food.*

Our vision:
- To promote organic produce and to set up a demonstration organic farm with high production in a small area.
- To use our farm to teach others how to grow fruit and vegetables organically.
- To earn over $100,000 p.a. from our produce selling as much as possible locally.
- To build a residential training centre on our property.

Our strategies:
- Leave the land more fertile and productive than it was when we purchased it by using a "tithing" system for our land, putting back 10% or more of the revenue into fertilizers and soil enriching minerals
- Set up a weekly farmers' market on Mt. Tamborine which benefits us and all local growers
- Market all our produce without advertising and with minimal cost and effort
- Educate ourselves
- Develop healthy balanced soil with regular ongoing fertilising
- Mulch heavily and regularly
- Develop a local organisation that supports the community and encourages self-sufficiency
- Practise bio-diversity
- Minimise the cost of labour and machinery
- Develop training courses teaching soil, plant and human health

Chapter 21
Action Plan

"Nothing happens unless first a dream."—Carl Sandburg.

This is a summary of the key action steps needed to get underway quickly and efficiently (the number in brackets is the chapter in this book which covers this activity).

1. **Choose property and decide layout. (5 & 6)**
The availability of water is the most important factor.

2. **Decide on farming method** (e.g. biodynamics, organic certification, nutrition farming). (2)
You have available to you the most amazing opportunity to learn if you care to use it. We had an Australian woofer stay with us who was about to retire. He had decided to retire to a country property and do some farming but knew nothing about farming. His excellent solution was to woof on a variety of farms to see what others did, including biodynamics and organic farms.

3. **Decide whether you need a shed &/or carport as a shelter and storage area.** (24)
We invested in building a custom-made packing shed close to the house. This has been very valuable both as a work area and a place to store fruit and vegetables, crates and fertilisers. It also provides an area to wash vegetables. We plan to build a "carport" to shelter our rideon and trailers which are currently exposed to the rain.

4. **Prepare a business plan**. Set some key goals. (19 & 20)
Without a business plan in our first six years we did not make much money. Clarifying the priorities made a huge difference to our income. I did not realise the significance of growing more fruit trees until we set out the revenue projections and started comparing the returns from different fruits.

5. **Do a soil test. Get a prescription blend fertiliser**. Order mulch.(3, 5,15)
Our soil pH was far too low in the first five years and it was not until we invested in a prescription blend fertiliser from NTS that we achieved the magic 6.4 pH. We could see the difference in the health and productivity improvement.

6. **Get some good tools & a rideon or tractor**. (24)
We had two secondhand rideons before we invested in a more powerful new rideon. The first two were continually breaking down, losing us time and money. This new machine has had no breakdowns so far and is faster and more comfortable (see photos 35 & 47).

7. **Set up a composting system.** (15)

Compost and mulch are essential to obtain lots of life in the soil and to keep the weeds down.

8. Decide which vegetables & herbs to start with. (9 & 11)
This gives short term cash flow. It took us a while to find out which are the most profitable. For example I was carried away by the price of garlic ($20 per kg wholesale and up to $40 per kg retail) before I realised the time it takes to plant, grow and harvest it. Carrots, kale, rhubarb and silver beet make much more money with far less hassle.

9. Prepare ground for vegetables. (15)
Time spent preparing the ground and planting with care is the foundation to productivity. It's like building a house—the foundations are what's important. The "**rule of seven**" is vital to successful growing: the first **seven** days for vegetables and the first **seven** weeks for trees. My neglect of our avocados in their first seven weeks has cost us thousands of dollars.

10. Put in irrigation and a fertigation unit for vegetable area. (16)
I strongly recommend putting in irrigation pipes above ground i.e. do not bury them. Maintenance is much more difficult if they are buried and you are less likely to damage the pipes with forks or spades if you can see the pipes.

11. Decide on at least three fruit trees, nuts or berries. (10 & 12).
If you want to achieve a goal similar to the one in this book's subtitle, this is the most important decision you will make. You will need to plant at least 300 fruit trees and possibly 500 for a five acre property. On a smaller property, say one acre, you will need at least 100 trees. This creates a lot of work and it is vital you assess the demand and ensure you have a good idea of how and where to sell the fruit. (22).

12. Calculate area required for fruit trees and decide on which area they will grow in. (23)
In chapter 23 I illustrate the calculations you need to make for each of the three key fruit I chose, namely avocados, blueberries and limes. In hotter climates consider mangoes or paw paws. In colder climates consider cherries and pears.

13. Order fruit tree seedlings. See Appendices C & D.
Quality is the key requirement as these fruit trees should last 20+ years and provide the bulk of your income.

14. Prepare ground for fruit trees. (15)
See point 9 above. If I had realised the importance of soil preparation and the need for attention to foliar sprays in the first seven weeks after planting I would be a lot wealthier today.

15. Put in irrigation and a second fertigation unit for fruit trees. (16)

If you put in the recommended 300+ fruit trees, I strongly recommend you invest in a fertigation unit to facilitate the fertilising process. See also point 10 above.

<u>Suggested timeframe—the above program will give you a great start and is worth taking one year to 18 months to give you a sound foundation for success.</u>

Steps 2 to 8 can be done in a month but time taken here is worthwhile.
Step 9 will take two to three months to do properly.
Step 10 should take only a couple of weeks.
Steps 11 to 13 can be done in a few days but time taken to research these issues will be repaid.
Step 14 should take 9 to 12 months.
Step 15 can be done anytime in the last three months of the time that step 14 takes.

Chapter 22
Marketing

"Life is painting a picture, not doing a sum."—Oliver Wendell Holmes.

Marketing is a major consideration for any successful business. Marketing is the 20% of what you do that gets 80% of your results. The decisions you make here are vital to the success of your venture. The good news is that they are easy to change. If marketing is not working you can readily change direction and there are many marketing consultants you can seek advice from.

Our business grew steadily. The major reason for our success is that we grow and sell quality food which is nutrient-rich. The alternative is commercially grown food which carries herbicides and pesticides or imported food. Our food is fresh. We generally pick produce today and sell today or tomorrow. Commercially grown food sold to supermarkets or through market agents generally takes 2 to 3 weeks plus to get to the customer. Imported food takes months. Since nutrient content decreases with time this gives our fruit and vegetables a huge advantage.

Furthermore we focus on ensuring all the natural elements are in the soil with high levels of organic matter, providing food for the microbes, worms etc. who convert the elements into plant food. It is a fact that fruit and vegetables grown on different soils have different nutrient content. If the soil is deficient in one or more elements, the plants grown on that soil cannot contain those elements.

Our marketing is different because it uses an educational approach to get its message across by promoting organic farming with its major benefits to human health and the environment. We aim at increasing public awareness of the link between good food and good health. We do this through our training courses and our newsletter. We also demonstrate that organic farming can be profitable. This approach increases the profile of organic produce and increases our own sales, as education increases awareness of the benefits of eating nutrient-rich food. It seems to work when you see the amount of free publicity we get.

First we made up boxes of fruit and vegetables and tried weekly deliveries to homes. (This is what FoodConnect in Brisbane does very successfully to about 1000 homes).

Next we set up a roadside stall at our front gate. Some roadside stalls do really well e.g. some friends sell most of their avocados from a stall fetching over $1000 a week for about 8 months of the year.

My wife then had one of the best ideas we have had, which gave our business a major leap forward. She stood up at a general meeting of our Local Producers Association and said: "Why is it that Tamborine Mountain successfully markets wine to tourists when it grows hardly any grapes? It has been promoted as a wine tour venue and we have six so-called wineries. Why don't we market what we actually grow to the thousands of tourists and visitors who come up here every weekend? Let's start a weekly farmers' market." There was considerable scepticism from many of the old hands: "We've tried that before, lassie, and it doesn't work." Well three months later we started up. We found a venue which we still use and we have opened for business almost every Sunday morning for over seven years now. The market turnover has grown steadily. We run it through the Local Producers' Association (LPA) which I have been president of for the past three years. Our market has become a community institution, a social event for local residents and an attraction for tourists and visitors. We have won awards and been featured regularly in newspapers, magazines and television. We have done almost no advertising—all we do is put out a few roadside signs and a couple of banners at the entrance, so that new people can find us.

The basic format of how we operate has not changed even though it is different from most farmers' markets. We do not have separate stallholders, we all work together and have one or two people sell on behalf of all suppliers. We agree on a common price for each product, so that if ten people want to sell avocados they all put them in bags and they will sell for $3 irrespective of how many are in a bag. If someone wants to sell more, they either put bigger avocados or more in each of their bags.

The aim is to maximise the return to the growers who receive 85% of the retail price and do not have to rent or man a stall. The LPA takes 15% to cover the rent, insurance and running costs which include a fee (paid as a rebate against the supplier's 15%) for those who do the selling for the others. Suppliers bring their produce at 7.00am with a sheet of paper saying what they have brought. At 12 noon they come back and get paid and take away any produce not sold. The produce is owned by the supplier until sold.
The LPA makes a profit from its 15% and returns this to those who supply produce in various ways such as subsidised fertilisers and a 50% subsidy on an annual soil test (to encourage people to do soil tests). We also usually have a subsidised Christmas Party each year.

Our format was copied by Beechmont (40 minutes drive south) last year and they now run a similar weekly market also on Sundays. We supply them with some produce and they can supply us with some of their items that they produce in larger quantities than can be sold at their market.

Our local farmers' market clearly gave us the sales outlet for our vegetables which we do not generally grow in sufficient quantities to warrant taking to the main Brisbane markets at Rocklea. We did initially take our avocados to Rocklea. We stopped doing this because of the packing and preparation required by them and the poor pricing and slow payment. They required us to dip our avocados in a toxic substance to kill any bugs on the surface. The substance was so toxic it was dangerous to handle. We also had to grade the avocados carefully and ensure they were clean and reject any with markings and then they had to be packed in special trays with plastic "plicks" to hold them in position and make them look good.

We also sell some avocados at our local market. The advantages of a local market are huge. There are savings in transport costs and packaging. The market is a cash business, so there is no invoicing or debtors to chase up. You get to deal direct with the consumers, so you learn what they want and on top of all that there is a great sense of community. You get appreciated for your efforts because people value the opportunity to buy local, fresh organic produce that tastes good.

Selling through a nearby farmers' market gives us an outlet for 30% to 50% of our produce, most of our vegetables and a few fruit. We sell also to the Beechmont market and to SongBirds, our local restaurant that also has cabins. The larger volume produce need one or more other outlets. We currently sell about 25% of our produce to FoodConnect; as our income from fruit grows we expect to sell 50% to FoodConnect, who are also growing.

What other ways do we market?

We have a website (www.greenshed.com.au) which is maintained for free by one of our suppliers. We publish a free monthly newsletter with recipes and nutritional information. We also know that we can sell at retail prices to local restaurants and make a delivery every Wednesday. Our friends at Beechmont market sell to their local school tuckshop. I had planned to set up a customer database, particularly for restaurants, say within 30km., and do a direct mail once a month with what's in season and a few special offers. I decided we would deliver free if the order was over $100. However we have had such a high demand at our market that we have not had to do the direct mail campaign and we have not had to make special offers. Occasionally towards the end of the market day if there is a product left over we offer it to latecomers and other suppliers at a discounted price.

Once we became disillusioned with the Rocklea market, we sold to a big farmers' market at Warwick about 180km. west of Tamborine Mountain for a while until we came across Robert Pekin, who started FoodConnect in Brisbane. Our organic wholesaler, FoodConnect, does not require special packaging, does not require grading and most important does not require dipping in a toxic substance. Moreover we generally get paid within 2 or 3 days. We were fortunate to be introduced to Robert Pekin by Graeme Sait at the time when Robert was starting up his Brisbane FoodConnect operation. He distributes weekly boxes of fresh organic fruit and vegetables to Brisbane families. His business has grown in parallel with ours and we have been delivering weekly to him for over three years.

Bev and I coordinate (through the TMLPA) sales from other suppliers to FoodConnect and often take up produce from them to Robert. After awhile we thought why should we return home from Brisbane with an empty vehicle, when Robert has nearly 100 organic growers in S.E.Qld. delivering produce to him every Monday and Tuesday. Even though we grow 50 to 60 different crops there are still many that we do not grow or that are in demand but out of season on Mt.Tamborine. For example, apples and pears, all the stonefruit and pineapples that do not grow well on Tamborine Mountain but that do grow elsewhere in S.E.Qld. So we agreed with Robert to buy some of this produce each week for our market. He orders an extra box or two from the other suppliers for the items we need regularly and the suppliers sell more and everyone is happy. This has helped the growth of our market by providing a greater range of produce to our customers. We also buy some produce from FoodConnect for the Beechmont market, which saves them time and money as they pick up from us and we are 40 minutes drive instead of over 2 hours if they went to Brisbane themselves.

To handle this extra volume we needed a coldroom to store the produce in and the TMLPA has been able to buy two secondhand coldrooms for this. As the coldrooms are owned by the TMLPA any member can also use the coldroom and this is also helpful for those suppliers for whom we sell produce to FoodConnect. They deliver it to the coldroom one or two days before we take it up to Brisbane.

If you need to advertise then there are lots of avenues that can help. TV is obviously effective but expensive and we are pleased to get the occasional free TV publicity. Local radio is much lower cost and an avenue we have thought we would use to boost our local market if we have the need. Publicity in the local press is good and you can run paid advertisements. Speaking at local events and to local organisations can help spread the word. Writing a book is an option.

Prices

We sell without the common premium for organic produce, because we buy much of our supplies at wholesale prices and sell most of our produce at retail pricing at our local farmers' market.

At the market we have a "standard" price which does not fluctuate depending on supply but generally stays the same all year round. Price reflects the cost of inputs: labour, preparation, growing time, skill required for cultivation, difficulty of picking. Price is set at a level judged to be "fair" to the grower.

Market Size

The Australian organic sector was estimated to be worth $400 million (Australian) dollars in 2005. Production has been increasing between 6 and 15% p.a., whereas consumption is growing at 25 to 40%. The difference is made up by imports. The organic sector is still small in relation to the total food market and that is an opportunity because the trend is upwards.

How is the advertising media changing?

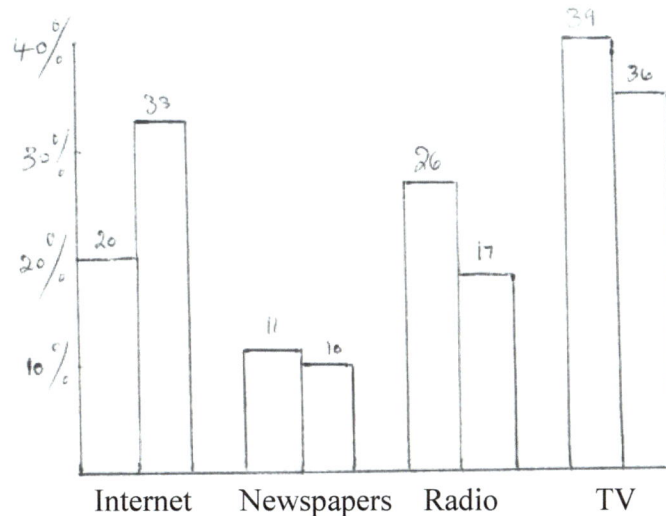

Bar chart showing percentages for Internet (20, 33), Newspapers (11, 10), Radio (26, 17), TV (39, 36). Y-axis marked 10%, 20%, 30%, 40%.

2002 compared to 2007.

Internet advertising is growing at over 25% p.a. We are seeing the steady decline of traditional media at the expense of the growing influence of the Internet. Newspapers and magazines are contributing to their own demise by raising prices when Internet advertising is much cheaper. The amazing statistic is that 79% of Australians are now regular users of the Internet.

The main promotion of our local farmers' market is by word of mouth and repeat business. The market does have its own website and we have also started our own website called http://healthyorganicnow.com which also offers our newsletter free and will also sell this book in a down loadable format. We are working to promote our organic farming and nutrition training courses through online marketing.

The advantages of online marketing

It is worldwide, low price or free, provides customers with instant gratification, has relatively low startup costs, is easily measurable (taking the guesswork out of advertising), can be fully automated and creates a level playing field (with big business).

There are some disadvantages compared to other media:
1. Penetration – TV is in 99% of households (in the western world) and the Internet is only in 62%.
2. Advertisements are small and have to compete with many others.

Like most things, there is expertise needed to do it well but there is no doubt your marketing expenditure will be lower if you use online marketing. The cost of a display advertisement in a newspaper or magazine can be over $1000 and what do TV advertisements cost? If you are not using internet marketing then your marketing budget is probably higher than it should be. One of the changes that the internet has brought about is still amazing to me. When I was in business (before I started farming), I specialised in direct mail and the cost of sending out mail was roughly $1 per envelope (postage, envelope, cost of printing the inserts etc.) not counting the labour. Nowadays you can send out an email campaign for less than one cent per item. What has also changed is the ability to reach people on a more customised, more personalised basis.

Chapter 23
Revenues and expenses

"If we all did the things we are capable of doing, we would literally astound ourselves."—Thomas Edison.

Here are the revenue projections and some actual revenues achieved and our plan to achieve the goal of $100,000 p.a. from five acres. For a smaller area, the figures are not proportional e.g. $30,000 p.a. from one acre (studies have shown the smaller the area the higher the productivity).

Trees—fruit and nuts.

a) Revenue from avocados.

Assumptions:

Spacing = 5 yards x 5 yards or 5 x 5 metres or 4 x 6 m.
5 m. spacing gives 12 x 12 trees = 144 per acre or 432 in 3 acres.

It is also possible to grow some vegetables (such as tomatoes or rhubarb) under the trees. Avocado trees are a wonderful investment. I know of one tree on Tamborine Mountain that is over 50 years old and is still producing over 1000 avocados a year, so do not stint on the initial cost. They should be planted from seedlings on good quality rootstock and can cost between $15 and $30 each. Do not plant them from seeds. Seeds can take 10 years before they fruit whereas seedlings have fruit in year three or even earlier.

Revenue targets:

400 trees @ 333 avocados per tree @ 75 cents per avocado = $100,000 p.a.
From our avocado study we know that it is **possible** to achieve an average of 449 avocados per tree. I suggest you aim at an average of 300 avocados per tree p.a. and this will take 4 to 5 years to achieve. I started with 300 and now have 250 trees and here are our revenue projections:

Avocados	2007	2008	2009	2010	2011	2012
230 mature trees	10000	12500	20000	25000	40000	50000
10 Hass 2/07	0	0	200	750	2000	3000
10 Fuerte 8/07	0	0	400	1000	2000	3000
Total no.of fruit	10000	12500	20600	26750	44000	56000
Revenue	$7500	$9375	$15450	$20062	$33000	$42000

N.B. 56,000 from 230 mature trees is an average of only 243 avos. per tree.

(b) Revenue from blueberries.

Each bush can produce up to 15 kg. but 5 kg. is more usual.
125gm.punnets sell for $5.50 at current prices (see photo 34).
If you get 5 kg.per tree, you receive $220 per tree.
At 10 kg.per tree, you would get $440.
300 trees @ say $200per tree = $60,000 and 300 trees @ $400 per tree = $120,000
p.a.

A key factor in the profitability of blueberries is the small space they take up, as they are planted only 1m. apart. I planted 320 and now have 280 bushes.
Each bush takes up 1 sq.m.so 280trees is an area 20m. x 14m.

Blueberries	2008	2009	2010	2011	2012	2013
100 plntd 2/07	0	2kg/tree	3kg/tree	5kg/tree	6kg/tree	7kg/tree
Revenue	0	$8800	$13200	$22000	$26400	$30800
180 plntd 6/08	0	0	2kg/tree	3kg/tree	5kg/tree	6kg/tree
Revenue	0	0	$15840	$23760	$39600	$47520
Totals	0	$8800	$29040	$45760	$66045	$78320

(c) Revenue from tamarillos.

Each tree can bear 10 to 20 kg. of fruit.
80 trees @ 10kg. = 800 kg @ $4 per kg. = $3,200 p.a.
Tamarillos normally bear fruit from February to July. With severe pruning around Christmas time, this delays fruiting till July/September when they are scarcer and you can get a much better price. This will clearly improve the return.

(d) Revenue from limes.

There is a steady demand for limes from restaurants and bars. Limes are also a less popular crop so the competition is not as great as for avocados where big business is starting to get in on the act and imports from New Zealand can affect prices significantly. In our climate limes start picking in mid-December and run through till July. Limes do well on Mt. Tamborine and are a small fruit which is a big plus for transport. To give an example of the revenue potential of limes, they yield about 9 kg. of fruit in year three but often exceed 100kg per tree by their 10th year. This would return 100 trees @ $3 per kg. @ 100kg. = $30,000 p.a. once mature. This could be conservative because any early fruit in December/January will fetch much higher prices and good trees can produce over 150 kg per tree. This is pretty good but will take 5 to 10 years to achieve. Limes can be grown slightly closer together being smaller trees than avocados and this is another advantage. I am investigating timing of pruning for limes as there is a suggestion that their season can be extended by judicious pruning. Lime saplings cost about $13 each.

Area required: 2.5 to 3m apart i.e. between 6.25 and 9 s.m.each or 100 take between 625 and 900s.m. or say an area 10 x 90 metres.

This projection for our property is based on planting 100 limes which we did in August/September 2008 returning $3.50 per kg.

Limes:

Year	1	2	3	4	5	6
	2008	2009	2010	2011	2012	2013
Prd'nfor 100 trees	0	100kg	900kg	2000kg	4000kg	6000kg
Revenue @ $3.50	0	$350	$3150	$7000	$14000	$21000

(d) Summary of fruit tree plan for our property.

250 avocado trees @ 300 per tree @ 75 cents = $56,250 p.a.
250 blueberries @ $250 per tree = $62,500 p.a.
80 tamarillos @ 10kg @ $4 = $3,200 p.a.
100 limes @ $3.50 per kg @ 9 kg = $3,150 p.a. (year three) growing to 100 x $3.50 x 100 kg = $35,000 p.a. by year 10.
Total for fruit in year three = $50,000 growing to $160,000 p.a. in year ten.
Macadamias will add to our income from about year five but I am not sure of the yield and pricing and have not included them in the above projections.

(e) Summary of potential income from fruit grown on our farm.

Financial year	2008/09	2009/10	2010/11	2011/12	2012/13	2013/14
Rev.from avos	$9375	$15450	$20062	$33000	$42000	$47625
Rev. from blues.	0	$8800	$29040	$45760	$66045	$78320
Rev.from tams.	$3200	$3200	$4000	$5000	$6000	$6000
Rev.from limes	$350	$3150	$7000	$14000	$21000	$24500
Totals	$12925	$30600	$60102	$97760	$135045	$156445

Revenue from limes and macadamias should continue to increase beyond 2013/2014, so you can see that 3 or 4 fruit tree combinations can take you way over $100,000 p.a. on five acres. Other citrus, paw paws, custard apples, mulberries and bananas generate only very small income for us.

Revenue from vegetables, herbs and berries.

Most of our early income came from sales of vegetables through our weekly market.

In 2006/2007 we started selling once a week to SongBirds, a local "organic" restaurant on the mountain. In 2007/2008 we started selling weekly to the Beechmont farmers' market.

In 2007/2008 we sold produce to the value of $30,000, an increase of 100% from 2004/2005.

We supplement our income in two ways: by buying in fruit and vegetables from FoodConnect and other sources at wholesale prices and reselling at retail at our local market, and secondly by training people in how to grow their own food. Both of these are related activities that would not be feasible without our farm.

Our largest income from our own vegetables has been from rhubarb, followed by kale, ginger, garlic, silver beet and carrots.

Rhubarb, kale and silver beet are special crops in that they can be picked several times a year from the one plant.

Expenses.

Seeds: we spend from $1300 to $1600 p.a. on seeds

Fertilisers: our expenditure on fertilisers has increased as we realise its importance. We generally now spend between $3000 and $4000 p.a. This will increase as we find that increased fertilising means increased productivity. One of the top farmers on Tamborine Mountain says you should spend 10% of the crop's revenue on fertiliser.

Mulch: we spend generally between $1500 and $2500 p.a.

Maintenance: equipment (largely tractor and rideon) and irrigation maintenance generally runs between $2000 and $2500 p.a. This includes diesel for the tractor and petrol for the rideon.

Our stationwagon is expensive in terms of maintenance, registration, insurance and petrol but a large proportion of this is personal use.

Electricity: our bill has risen steadily because of price and usage increases. We have nearly 700 sprinklers operating which use our bore pump and we put in a coldroom which operates 24 hours a day @ 4 degrees C. To combat this increased cost we are implementing four strategies: (i) solar panels to generate our own electricity and create a credit by selling the surplus to the grid at a high price; (ii) moving to a two tariff system so that at night we get a much lower cost affecting our coldroom usage and our irrigation which will switch to night time; (iii) putting in insulation including a sort of carport over the coldroom so that it is not directly exposed to the sun; (iv) putting in a solar hot water system.

Summary.

You will see from the projections of revenue above why I am confident that our revenue will continue to grow, as in time blueberries, limes and macadamias will add another $100,000 p.a. from just these three crops. In my case I planted them too late to achieve the goal within five years. In your case I recommend you plant avocados, blueberries and limes (or other fruit trees of your choice) by the end of year one and you can achieve results much quicker than we have. I recommend you install a shadehouse at the outset and treat it as part of your initial capital investment. This will further improve your productivity particularly when it's wet.

We have planted more rhubarb each year and expect our income from our own vegetables to reach $35,000 p.a. **So the mix of produce to generate $100,000 p.a. solely from our own produce will be about $65,000 from four fruit types and $35,000 from about 50 vegetables. On one acre you could expect $10,000 to $15,000 p.a. from vegetables and about $20,000 p.a. from fruit trees.**

Chapter 24
Investment Required

"Anyone who lives within his means suffers from a lack of imagination."—Lionel Stander.

<u>**Equipment needed:**</u>

a. Vehicle for transporting produce. We get by with a stationwagon and a trailer.
b. A ride-on or zeroturn with trailer &/or tractor with slasher & trailer.
c. Spray equipment.
d. Irrigation (pipes, control unit, sprinklers) plus a fertigation unit.
e. Tools such as chainsaw for pruning etc and crates for packing
f. Cold room.
g. Optional – a spreader.
h. A shed for packing, storing fertilisers & crates/boxes. The shed ideally needs basins/water supply for washing vegetables.
i. A shadehouse or greenhouse.

You need a vehicle for transporting produce. In our case this doubles for our normal personal driving use. If we get much bigger, we may have to get a van, but up till now we get by using the towbar and a trailer. We extended the use of the trailer by getting a wire "cage" made for it which enables us to transport much more with safety. After ten years we are onto our second trailer. They take some skill to park and to improve their lifespan you ideally need to be able to park them in an undercover area that's easy to back into (or a "drive thru" carport), so that they do not rust in the rain. We added a fitted cover to our cage recently to protect the produce from wind and rain (it also stops empty boxes flying out the top). (See photo 32).

You do not have to invest in all the above items at the outset. The next chapter records our timetable for when we bought most of the items.

Costs and personal preferences will vary a great deal. Our current trailer cost $1941 in 2007 including registration for onroad use. The smaller trailer for use with the rideon cost only $399 but is one of the most used items on our farm. We seem to be forever moving things: mulch, fertiliser, produce, rubbish……… Our two fertigation units, one for the fruit trees and one for the vegetables, each cost $2000 to install (see photo 33).

Our second chainsaw was a bigger one and cost $899 in 2008. It gets a lot of use and the longer blade makes it much more effective, speeding up the process.

Our biggest investment was to build a custom-made packing shed in brick and tile (to match our house) and this cost $50,000 in 2005. It has a large work room and an open area at the back with three sinks for washing vegetables (see photo 41).

Other "investments".

An item which is usually expensed but which requires a cash outlay is seeds and seedlings. Seeds can be very expensive and we buy them in bulk to reduce the cost and have learned to keep seeds in the refrigerator or coldroom. Seeds are mainly for vegetables which have a short lifecycle so should be treated as an annual expense item. But seedlings are an even bigger investment and are a lifetime investment and so could be capitalised. Top quality avocado seedlings can cost up to $30 each, so 300 require an outlay of $9000. I bought 200 blueberries this year (2008) for $600 and 100 lime trees for $1350. We decided to put some sheep in to do the weeding and mowing under and around the avocados. Ten sheep cost us $700 and our ram cost $350. A bigger cost was the fencing and gates to keep them in and these cost us $3000.

A coldroom may be needed to store fruit and vegetables to protect them from the heat. We have been fortunate that our Local Producers" Association has invested in two coldrooms, which are available to suppliers to our farmers' market and for produce to be delivered to FoodConnect in Brisbane. Both were acquired second-hand and both needed new compressors. They also draw a fair amount of power each time the compressor kicks in, and so electrical installation costs can be significant. Both of ours cost $4000 to become fully operational.

We are onto our third ride-on mower and general workhorse. The first two were secondhand and 13 to 15 horsepower. With our slopes and clay soil they got a good workout and this year we invested in a new John Deere ride-on that has 25 h.p. This cost us $9300 (see photo 43). We are hoping to sell our Cox ride-on (see photo 37) to offset some of this cost but we will be lucky to get $2000 for it.

Spray equipment is another outlay. I started with a backpack for less than $200. I then graduated to a spray unit that fits in the trailer behind the rideon and runs off the battery of the rideon. This unit cost about $700 (see photo 34). A better solution was to get a bigger unit that fits on the back of the tractor and runs off the PTO unit which gives it more power. This unit takes 300 to 400 litres. I borrowed this spray unit from a friend as the cost of a new one can be at least $2000 and they do not get a lot of use. I have been looking for a secondhand one without success and as mentioned elsewhere I have recently decided I can afford to pay a contractor to do this job for me.

We acquired our Kubota Diesel tractor with the house and property (see photo 31), and have had it for over 12 years. It is probably at least 20 years old but we were fortunate to find an expert person to maintain it. He loves it and says it will last forever. It is very reliable and we get it serviced only once a year. When we bought our John Deere ride-on we had decided we needed a more powerful rideon than before. They come in many sizes and the bigger ones would have obsoleted our tractor. We made the decision to get a more powerful smaller one rather than a bigger one, so that we still had a rideon and a tractor. Clearly if one breaks down you then have a backup machine, because you always need a machine to move things. Also the rideon is much more manoeuvrable in tight spaces and can get closer to fruit trees when mowing than the tractor. The tractor has a bigger span for mowing and is more powerful. Two machines also mean two people can be doing some work at the same time. We spent $1200 on a new slasher for our tractor.

You may need to build a shelter, garage or a carport to protect your machinery as rain and rust shortens their life. Our house is built on a slope and has two garages, one for the car and one for the tractor. We do have a problem with storing the rideon and trailers out of the rain.

Another possible area of investment is a dam, depending on your water situation. We are fortunate to have a bore that supports 600 sprinklers. We did put a dam in last year but the reasons were more for adding to the beauty of the place, providing a swimming hole for our woofers and as a water source to comply with some fire management guidelines (see photo 4). To build the dam wall and create the dam including a bridge over the stream cost $12000 in late 2007. We also put in a pump to recycle the water from the dam back upstream, so that when the stream dries up we can make it flow again. Flowing water adds something to the beauty and spirit. This pump also stops the water in the dam from becoming stagnant.

A shadehouse can cost between $5000 and $25000 depending on size, type and fittings. We have not put one in yet. A friend is planning to put one in shortly and in subsequent editions of this book I will report on his experience and maybe mine if we go ahead with one.

A summary of our total capital outlays over several years shows:

Packing Shed	50,000
Machinery	18,400
Seedlings	5750
Sheep and fencing	4000
Dam	12000

Total	$90,150

Chapter 25
Diary of what we did and when.

"We must always change, renew, rejuvenate ourselves; otherwise we harden."—Goethe.

1995
Bought our property.

1998
Planted 300 avocados
Started growing vegetables

1999
Started selling produce to organic fruit and vegetable shops
Planted 20 avocados

2000
Started selling boxes of our own produce and set up a roadside stall

2001
Started local farmers' market
Doubled vegetable growing area
Obtained an ABN from the Taxation Office

2004
Our vegetable garden was chosen as one of the Tamborine Mountain "Open Gardens" for the public to visit in the Springtime Festival.
We again doubled our vegetable growing area
Did a "Certificate in sustainable agriculture" with Nutri-Tech Solutions P/L

2005
Built a packing shed (August/September)
Planted 25 macadamias
Installed fertigation system for avocados
Started teaching "Growing Healthy" program
Put prescription blend fertiliser from NTS on avocados

2006
Started selling avocados to FoodConnect
Did a fertigation of avocados with beneficial fungi spores

2007
February- planted 120 blueberry plants and 10 Hass (on Dusa clonal rootstock) one year old avocados
April—installed a secondhand coldroom.
July--- planted 10 finger limes and 10 Fuerte (on Dusa clonal rootstock) two year old avocados
July –bought new trailer for car & tractor
August—acquired a moveable chicken coop
September--installed a fertigation system for vegetables
September – won Small Business Champion award for Queensland for Fresh Food
November—satellite local farmers' market started at Beechmont
December—built a dam on stream

2008
January-installed a fence to keep sheep in.
March –12 sheep arrived.
April—acquired a ram (see photo 42).
April—registered business name "Growing Healthy"
May—took on our first paid employee
May –finalist in the Australian Vegetable Grower of the Year awards
June—planted 200 blueberries (see photo 29).
July –bought a new John Deere rideon with 25 hp.
July—planted 40 tamarillos
August—planted 100 lime trees (see photos 15 & 17).
September— our first lambs were born (see photo 44).

With some of these steps there comes a lot of planning. For example, with the planting of new trees, there are several other key additional steps in the preparation process. Specifically with the 100 lime trees the steps were:

1. Ordered 100 trees in August 2007.
2. Decided where to plant them in April/May 2008.
3. Prepared soil in April/June 2008 .
Ideally the soil preparation should have been started much earlier, in August 2007.
4. Obtain 100 stakes; put in irrigation system; obtain 100 sprinklers.
5. Planted in August/Sept. 2008.

Future plans:

2009---
   • Build training centre and market training programs.

- Take on more employees.
- Build chicken enclosure in orchard and install moveable chicken coop and an incubator to breed chickens.

2010---

- Build a carport to shelter rideon and trailers
- Install a shadehouse to protect against winds and hail and the extremes of heat/cold to produce tomatoes and provide a place for woofers and staff to work when it rains.
- Acquire chicken for our own eggs and meat and to keep the grass and weeds down around the fruit trees.

One of our early actions was to build a chicken coop. We thought producing our own eggs was part of the goal to be self-sufficient. We had about 16 chicken for 3 or 4 years, but eventually gave this activity away for two reasons. One was the fact that chickens need constant daily attention, to collect the eggs and provide feed and water. Early on we liked travelling and this created a problem. Secondly we had an ongoing battle with a python and a goanna over who got to the eggs first. Now we are returning to this plan and intend to try having chicken again. We were going to get chicken in moveable coops to help weed under the avocados, but then decided that sheep would be a better solution for this problem. We now plan to have a couple of moveable coops in our fruit tree orchard with a fence around the orchard so that the chicken can roam during the day. Chicken are part of the sustainability cycle, eating scraps, providing manure, eggs and meat.

Chapter 26
Human health.

"May your garden bring much joy and satisfaction, and be a beautiful and interesting place, providing mental and physical therapy and a bountiful harvest for health and vitality."
Isabell Shipard, author of "How can I use herbs in my daily life?"

After 10 years of farming I have developed a totally new view of farming and farmers. Farming is far more intellectually stimulating than I had thought before. It involves you mentally, physically and spiritually. Working close to nature is uplifting to the spirit. I have never been happier and I am healthier and fitter for the physical work, after over 30 years sitting behind a desk. You cannot help but become healthier when you are working outdoors in the fresh air and sunlight away from the pollution of cities.

One of the important lessons I have learned from ten years farming is the strong correlation between soil health and human health. By way of illustration, a couple of years ago I had a series of kidney stones. The first bout was so painful I went to hospital and had an x-ray, cat scan and blood analysis. After eight agonising hours the pain suddenly stopped and the doctor came in and said you had a kidney stone and it has now passed into the bladder. You can go home now; it won't occur again. Over the next six months I had about seven reoccurrences. I saw another doctor and a herbalist and found many "cures" on the internet. None helped. What I did learn was that kidney stones are made of calcium and it suddenly occurred to me that the most important ratio for a balanced healthy soil was the calcium/magnesium ratio. I reduced my calcium intake and started taking a magnesium pill daily. From then on I have not had another kidney stone.

As I mentioned back in chapter one, there is an emerging world trend that recognises the value for all families and individuals to grow their own food. Russia is one country where this trend has taken off in a big way. Cuba adopted it as major national initiative when they faced the oil embargo. In 2009 Michelle Obama announced that the White House was starting its own vegetable garden.

Only when you grow your own food can you ensure the food you eat is both fresh and nutrient-rich and that it does not contain toxic chemicals. This trend will have huge impacts in saving money and improving health, as the primary cause of health problems is a deficiency in minerals in our diet. I believe for our own health as for healthy soils it is essential to have all the trace elements. As we have learned different plants have different mineral contents, so variety in your diet is important and to be absolutely certain you are getting all the elements you also need to take a good mineral supplement daily (see Appendix F for the results of my research on supplements).

In conclusion, I am now proud to call myself a farmer and I can truly say "I love what I do." I love our property and it is very satisfying to get up early and see the sunrise and the mists over the valley and the peace and calm at the beginning of each new day. "The farmer's goal is health—his land's health, his own, his family's, his community's, his country's" (Wendell Berry, 1977). I would add that the farmer's goal can also include the health of mankind and the planet.

I hope this book will encourage and help you to move towards growing healthy fruit and vegetables to help you grow healthier and happier.

Abbreviations

ABN = Australian Business Number

ABS = Australian Bureau of Statistics

ATO = Australian Taxation Office

ATP = Adenosine Tri-Phosphate (see Phosphorus in Appendix A)

CEC = Cation Exchange Capacity (see chapter 5)

CSA = Certificate in Sustainable Agriculture

DOA = USA Department of Agriculture

DPI = Queensland Department Of Primary Industries

EAL = Environmental Analysis Laboratory

GST = Goods and Services Tax

LPA = Local Producers Association

OM = Organic Matter

ORAC = Oxygen Radical Absorbance Capacity (see chapter 8 and appendix B)

ppm = parts per million

PTO = power take-off

TMLPA = Tamborine Mountain Local Producers Association Inc.

References.

Chapter 1.
1. "The Chemical Maze" by Bill Statham. Possibility.com.2006.
2. "New Nutrition" by Dr. Willem by Serfontein.2001 by Tafelberg.
3. "Animal, vegetable, miracle" by Barbara Kingsolver (2007).

Chapter 2.
1. "Priority One" by Allan J. Yeomans. Keyline Publishing.2005.
Chapter 3.
The comment on the three pillars of sustainable agriculture is based on Peter Andrews' book "Back from the Brink"—see pages 88 and 102.

1. Dept.Zoology and Endocrinal Laboratory, University of Wisconsin.
2. "Feel Better Live Longer" by Willem Serfontein, pages 24 & 25,published in 2005 by Tafelberg.
3. "New Nutrition" by Willem Serfontein, page 170, published in 2001 by Tafelberg.
4. "Minerals for the Genetic Code" by Charles Walters, pages 9 & 98, published in 2006 by Acres USA.
5. Jen Ross, "Paying the Price for Growth," The Toronto Star, Jan 8, 2005.

Chapter 5.

The section on CEC is based on notes from the 'Mineral Management' section of the Nutri-Tech Solutions Pty Ltd (NTS) Certificate of Sustainable Agriculture course and is printed with permission from Graeme Sait, CEO of NTS.

Chapter 6.

1. "Making your small farm profitable" by Ron Macher—page 89. Storey Publishing.1999.
2. "Earth User's Guide to Permaculture" by Rosemary Morrow (published in 1993 by Kangaroo Press).
3. "Vegetables Australia" magazine, Sept/Oct 2008 issue pages 46/47.

Chapter 9.
1. "The revolution will not be microwaved" (pages 45 & 54) by Sandor Katz (2006) Chelsea Green Publishing.
2. "The Seed Savers' Handbook for Australia and New Zealand." Byron Bay, NSW. The Seed Savers Network. www.seedsavers.net
3. "Breed your own vegetable varieties: the gardener's and farmer's guide to plant breeding and seed saving," by Carol Deppe. Chelsea Green Publishing.

Chapter 10.

1. "The revolution will not be microwaved" by Sandor Katz, 2006, Chelsea Publishing (page 33).
2. "The Biological Farmer" by Gary Zimmer (Chapter 19).

Chapter 11.

1. Minerals for the Genetic Code by Charles Walters (pages 79 & 111). Published by Acres USA in 2006. See also www.pubmed.gov under oncology and "alternative care" and "multiple remedies".
2. "Scandium and Yttrium in the Biological Process" by Charles T. Horowitz.

Chapter 13.

1. A large part of the material in this chapter is based on content from the "Certificate of Sustainable Agriculture" Course presented by Graeme Sait of Nutri-Tech Solutions P/L at Yandina in Queensland.
2. "The Biological Farmer" by Gary Zimmer—page 68. Acres USA. 2000.

Chapter 15.

1. "Back from the Brink" by Peter Andrews.
2. The Biological Farmer by Gary F. Zimmer—page 145.

Chapter 16.

1. "Avocado Grower's Handbook" published by the Qld. Dept. of Primary Industries.
2. "Back from the Brink" by Peter Andrews.
3. "The complete book of fruit growing in Australia" by Louis Glowinski. Published by Lothian, 1991.

Chapter 17.

1. "We want real food" by Graham Harvey.

Appendix F.

1. See article "Minerals from the Sea" by Linda Page, N.D., Ph.D. in Conscious Living magazine issue 7.

Bibliography

1. Take control of your health by Elaine Hollingsworth.
2. The Australian Fruit & Vegetable Garden by Clive Blazey and Jane Varkulevicius
3 Science in Agriculture by Dr. Arden Andersen
4. Back from the brink by Peter Andrews
5. The Upside of Down by Thomas Homer-Dixon.
6. Animal, vegetable, miracle by Barbara Kingsolver (2007).
7. The Bio-Gardener's Bible by Lee Fryer
8. The Biological Farmer by Gary F. Zimmer

Appendix A

The 18 key natural elements.

1. Aluminium (Al)

Al is the most common metal—more common than iron. However it is only needed at low levels in the soil, generally 5 ppm is adequate. Boron is easily displaced by high Al levels. In fact Al is very toxic at higher levels and in humans is suspected of being a major contributor to Alzheimer's disease (funny that this disease' name starts with Al!). The only antidote for Al toxicity is calcium, magnesium and silicon.

2. Boron (B)

Boron is necessary for the photosynthesis process which is covered in detail in chapter 14. B also governs the uptake and efficient use of calcium. It is essential for cell subdivision and development, it affects pollination, promotes flowering and is involved in the synthesis of proteins and hormones. B is easily leached from the soils. It therefore needs to be replaced at least annually. It is important to do a boron foliar spray prior to flowering. Boron in too large a quantity can be toxic and can kill plants; it is only required in 1 to 3 parts per million (ppm). Boron toxicity can be countered by the application of nitrogen or lime. Calcium and silica are both important plant strengtheners, but both require B to deliver. Plants need a constant supply of silic acid and B encourages silica to form silic acid.

3. Calcium (Ca)

Calcium (Ca) comes in several forms. The most common are lime, dolomite and gypsum. Dolomite is generally not a good source as it is high in magnesium and many soils are already high in magnesium. The calcium/magnesium ratio is one of the keys to good soil balance. Gypsum is good when preparing the soil—as it helps breakdown clay and improves the drainage. Gypsum has also been shown to reduce the risk of phytophera (a harmful fungus, commonly known as root rot).

Ca has been called "The Prince of Elements: by Dr. William Albrecht. It is the most important mineral in many chemical, biological and physical processes. It helps transport all minerals into plants. It is a component of every living cell. For a good balanced soil it is essential to get the right balance of Ca. You need to aim at a pH of 6.4 in your soil and this becomes self-adjusting when Ca, magnesium (Mg), Potassium (K) and Sodium (Na) are in the right balance. To determine this you need to know and check the key ratios. **The most important ratio is Ca/Mg** which ideally should be 7:1 in heavy soils and 4:1 in light soils. This ratio optimises soil structure, nutrient availability and high production. Ca helps ensure good levels of oxygen, the most important element in terms of microbe health. Lots of dandelions are an indicator of Ca deficiency in the soil. Ca levels needed increase with increasing CEC levels and are from 2000 to 5000 ppm.

4. Cobalt (Co)

Cobalt is another nutrient that is essential to fix atmospheric nitrogen. It also stimulates beneficial bacteria in the soil. It is required at only about .5 ppm but is another undervalued trace mineral. The combination cobalt-nitrogen-carbon is vitamin B-12. Co is a carrier and directs the use of ten other trace minerals: titanium, vanadium, chromium, manganese, iron, nickel, copper, zinc, gallium and germanium.

5. Copper (Cu)

Only about 5 to 10 ppm are required in the soil, but like many trace elements Cu is very important. It plays a role in many enzyme systems, is vital for effective root development and is also required for chlorophyll production and photosynthesis. Copper is a natural fungicide.

6. Iodine (I)

`Iodine is another trace element with an important role. It regulates potassium. Its function is to move potassium. It is very rare in Australian soils, so it is important to use fertilisers with sea salts or seaweed both of which have iodine. Because it is rare in Australia, a source needs to be found for human health as iodine is vital for the brain and thyroid functioning. Natural sources are kelp, pistachios and cinnamon.

7. Iron (Fe)

Iron is one of the most abundant elements but soils high in iron often do not make it available to plants. Fe is essential for the manufacture of chlorophyll and in the transport of chloroplasts i.e. it is vital for photosynthesis. It is also necessary for biological nitrogen fixation and a component of many enzymes. It is important for healthy leaves. If there is a deficiency of iron in plants, it is usually not because of a shortage of iron in the soil but because of an imbalance and rather than adding Fe it is usually better to work on improving the soil balance. It is less available at high pHs. Its availability is often related to sulphur and manganese. You need slightly higher amounts of Fe than manganese for both elements to be at maximum availability. Adequate sulphur usually means adequate iron. Alternatively a foliar spray is a quick way to fix deficiencies.

8. Magnesium (Mg)

Mg is the central element in the formation of chlorophyll which converts sunlight into energy by the process of photosynthesis. The Mg atom is at the centre of a chlorophyll molecule and plugs all life directly into the sun's energy. Any limiting factor that inhibits photosynthesis will limit productivity. So you need to ensure luxury levels of Mg in leaf analysis for both photosynthesis and for phosphorus uptake (the energy system of plants). Mg also plays a major role in seed germination and the synthesis of amino acids, vitamins sugar and several other processes. **High levels of potassium can tie up magnesium.**

Mg must be in the right ratio with calcium for uptake and soil structure (see above under Ca). Mg is easily leached so it is important to monitor levels both in the soil and particularly in the leaves (because of its critical importance in photosynthesis). The **second most important ratio for good soil balance is the Mg/K ratio** which should be close to 1:1, commonly from 150 to 400 ppm each.

Mg in humans and animals is important in the skeleton which holds 70% of the body's Mg. It is also necessary for the metabolism of fats and carbohydrates. In the USA 85% of people are deficient in Mg.

9. Manganese (Mn)

Mn accelerates seed germination and early maturity of crops. It has a role in the photosynthesis process. It has magnetic properties and electrically charges the seed, which in turn attracts other nutrients to the seed. Mn is also required for carbohydrate and nitrogen metabolism and is involved in many enzymes. Mn is required at 20 to 100 ppm in the soil.

10. Molybdenum (Mo)

Mo is the least abundant of the important trace minerals and is only required in the soil at less than one ppm. However it is essential for nitrogen assimilation, protein synthesis and production of the key reductase enzyme which is responsible for reducing nitrates to ammonium. High nitrates in a plant attract insect pests. Because of the high nitrogen use in agriculture it is important to ensure Mo is present to effectively metabolise the nitrogen. Without Mo nitrates build up in plant tissue, and high nitrates in our food deoxygenates our blood leading to many negative effects on the human body.

11.Nitrogen (N)

Nitrogen is used by plants more than any other plant foods. N is of course not an earth mineral but is supplied naturally for free by nitrogen fixation by soil organisms and from the decomposition of organic matter (OM) (see discussion of the nitrogen cycle in chapter 14). When supplied by man-made fertilisers it is very expensive. The carbon to nitrogen ratio is important and should be 10 or 11 to 1 (humus is generally 58% carbon).The other ratio of importance is the nitrogen: potassium ratio in the leaf which should be 1:1.

12. Phosphorus (P)

P is the key to yield and quality. All life processes require energy and P is the source of the plant's energy. P is the key element in ATP (adenosine tri-phosphate) which transmits energy in all living cells and is the energy that converts photosynthetic energy into glucose then complex sugars then starch. Energy in the form of ATP and sugars is required for pollination and vigorous shoot growth. A peak energy demand comes when plants form seeds and P is required at luxury levels for this to happen effectively. Also a healthy plant with lots of energy reserves has a strong immune system that protects plants from pests and diseases. P deficiencies often show up in stunted root and shoot systems and yellowing of leaves. Regular leaf analysis is the best way to identify P deficiencies. P does not leach. Mg has a strong impact on P uptake. So if you have poor P uptake in leaf analyses, by lifting the intake of Mg you will increase the uptake of P. One important ratio for P is the zinc ratio which should be about 10:1. This relates to leaf size and plant sugar production—each of these elements is capable of retarding the availability of the other if the 10:1 ratio is not maintained. P levels should be about 50 to 70 ppm.

13. Potassium (K)

Potassium is involved in nearly all aspects of plant growth. Growth requires a series of chemical changes. These changes require catalysts which are referred to as enzymes. K is a key element in about 50 plant enzymes. Some of its important functions are it is necessary convert nitrogen into protein, helps in the movement of sugars and enhances oxygen uptake for photosynthesis. K is second only to nitrogen in terms of quantity taken up by plants and is particularly critical during early growth. In some plants (such as bananas and avocados) potassium exceeds nitrogen. Most K in the soil is insoluble and needs to be released by weathering or microbial activity. Desired levels in the soil increase as the CEC level increases and are from 200 to 800 ppm.
Best sources of K are potassium sulphate or sulphate of potash and kelp. K is depleted from the soil by leaching. Sodium should never exceed K in base saturation as sodium will be taken up by plants instead.
There are some important ratios involving K for a good healthy soil balance:
Mg to K ratio should be close to 1:1.
K to Na ratio should be 5:1.
Ca to K ratio should be 15:1.
N:K ratio in the leaf should be 1:1.

14. Selenium (Se)

Se is another essential trace element that is rarely found in Australian soils. In humans it has been shown that Se helps combat cancer and the AIDS virus. It is considered by some to be the most important of all agricultural minerals as well as being involved in human's longevity. It is involved in many enzymes. Se has an impact on oxygen, carbon and nitrogen. It is only required at about 1 to 2 ppm. It is a good nutrient to add to foliar sprays. It is obtained from kelp and molasses.

15. Silica (Si)

Silica is the second most common element found on earth. It can play a major role in mineral uptake. It needs to be at least 100 ppm in your soil. Benefits of silica are:
- Increased photosynthesis from improved leaf strength.
- Absorption and translocation of nutrients within plants.
- Greater stress tolerance due to increased cell strength.
- Helps increase tolerance to salinity and heavy metals.
- Pest resistance.

16. Sodium (Na)

High sodium in soils destroys soil structure, and some crops are harmed by high sodium. Levels need to be about 40 to 120 ppm. The ratio of potassium (K) to sodium is important and should be about 5:1.

17. Sulfur(S)

Sulfur should be about 25 to 50 ppm in your soil and is often deficient. S is essential for the formation of chlorophyll and a key component of amino acids, stimulates seed production and helps seedlings survive, promotes rapid root development and is involved in the physical structure of plants. When plants are deficient in S, it compromises the plant's ability to transfer sugars and starch to the roots. An S deficiency usually shows up as a protein deficiency. S gets leached easily so needs replacing every year. Most of the soil's S is present in OM. S needs to be provided as sulphate and comes in gypsum and potassium sulphate. It has to be made into plant available form by bacteria. The sulphur to nitrogen ratio is very important, as sulphur improves nitrogen availability. This S:N ratio should be 1:10.

18. Zinc (Zn)

Zinc is critical as an energy micro nutrient and in the formation of growth hormones. One of its key functions is in the formation of chlorophyll. Leaves are the solar panels of plants and collect sunlight to convert into energy. Zn is a vital trace element and plays an essential role in many plant functions. A lack of Zn produces more serious symptoms than any other trace element. When Zn is deficient, applications of Zn have been shown to have the highest cost-to-benefit ratio of any nutrient applied (particularly as it increases photosynthesis).

Zn is a component of many enzymes, it helps in the absorption of water, it regulates the consumption of sugar, and it plays a role in carbohydrate metabolism. Zn deficiency means small leaves and stunted growth. You only need 5 to 10 ppm in the soil, but as mentioned above the P:Zn ratio of about 10:1 is important, so if P is high then Zn should be maintained at luxury levels to compensate. Zn is a very mobile nutrient and needs microbial activity particularly from fungi to make it available to plants.

Appendix B

ORAC Values of fruits and vegetables
(per 100 grams or 3.5 ounces)

FRUITS & HERBS

Cinnamon	267,536
Dried Goji Berries	25,000
Mangosteen	17,000
Dark chocolate	13,500
Artichoke hearts	7904
Wolfberry	7228
Prunes	5770
Raisins	2830
Blueberries	2,400
Blackberries	2036
Cranberries	1750
Strawberries	1540
Pomegranates	1245
Raspberries	1220
Plums	949
Oranges	750
Red grapes	739
Cherries	670
Kiwi fruit	610
Pink grapefruit	495
White grapes	460
Cantaloupe	250
Banana	210
Apple	207
Apricot	175
Peach	170
Pear	110
Watermelon	100

VEGETABLES

Kale	1770
Spinach,raw	1260
Brussel sprouts	980
Alfalfa sprouts	930
Spinach,steamed	909
Broccoli florets	890
Beets	841

Red bell pepper	713
Onion	450
Corn	400
Eggplant	390
Cauliflower	377
Peas,frozen	364
White potatoes	313
Sweet potatoes	301
Carrots	207
String beans	201
Zucchini	176
Yellow squash	150

References: 1.Wikipedia
2.www.drdavidwilliams.com/nc/ORAC_values.asp

Appendix C—List of seed and fertiliser suppliers in Australia

1.Seeds.

    (a) South Pacific Seeds P/L, Griffith, NSW. Tel. (02) 6962 7333

    (b) Yates Tel. (02) 9725 1088

    (c) Eden Seeds Tel. (07) 5533 1108 www.edenseeds.com.au

    (d) Norco Rural Stores @ Beenleigh Tel.(07) 3287 8999

    (e) Fairbanks—tel.(03)96894500—fax (03)96877089

    (f) GreenHarvest Organic Garden Supplies @ Maleny. Tel.5435 2699 or
        www.greenharvest.com.au

2.Fertilisers

    (a) Norco @ Beenleigh Tel. (07) 3287 8999

    (b) Daws for chicken manure by the bag 0413 602324

    (c) Nutri-Tech Solutions, 7 Harvest Rd., Yandina, Qld.4561.Tel 07 5472 9900.

    (d) BioFlora Ag Pty.Ltd. @ Beaudesert and Chandler.

    (e) For chicken manure by the truckload Arthy Rural Services Tel. (07) 5544 2266

Appendix D

Accredited Avocado Nurseries in Australia

1. Birdwood Nursery
   Peter & Sandra Young
   71-83 Blackall Range Road,
   Nambour, Qld
   Tel. 07 5442 1611

2. Anderson's Nursery
   Graham & Vivienne Anderson
   Duranbah Road,
   Duranbah, NSW
   Tel. 02 6677 7229

3. Avocado Coast Nursery
   Greg Hopper
   Schulz Road,
   Woombye, Qld.
   Tel. 07 5442 2424

4. Turkinje Nursery
   Peter & Pam Lavers
   100 Henry Hannam Drive,
   Walkamin, Qld.
   Tel. 0419 781 723

Appendix E

Certificate of Sustainable Agriculture Course

The course is run in Australia by Nutri-Tech Solutions (NTS) usually three times a year a
at 7 Harvest Road, Yandina, Qld. 4561. Tel. 07 5472 9900
info@nutri-tech.com.au www.nutri-tech.com.au
Cost: Aus $599 (incl.GST) (cost of a second person $299).
This course provides guidelines for high-production agriculture & human nutrition.
Knowledge is power, and in any business enterprise education confers the power to be
profitable. There are now technologies available which can ensure both increased
profitability and improved sustainability in agriculture.

The NTS Certificate in Sustainable Agriculture consists of four eight-hour modules on
soil and plant health interspersed with comprehensive information about human health:
Mineral Management
Microbe Management
Plant Management
Pest Management
Human Health Management
There is now generally a fifth optional day with visits to nearby farms.

You will learn the most valuable tips, tricks and synergies with this in-depth training
program which includes practical hands-on workshops.

The course will be presented by NTS' CEO, Graeme Sait, supported by a team of leading
biological agronomists. You will discover fascinating insights into proactive management
of both your soil and your own health.

Nutri-Tech Solutions (NTS), a world leader in soil, animal and human health, is a
dynamic, rapidly expanding company based on Australia's Sunshine Coast. NTS was
founded in 1994 to highlight the profound link between soil health and human health. We
are in the midst of a plague of degenerative diseases (including cancer and heart disease)
unlike anything the world has seen before. Many of these problems can be linked to
inadequate nutrition from our food and the associated lack of minerals and microbes in
our soils. This innovative company offers a complete, holistic approach including
education, state-of-the-art products and expert consultancy.

There is a tremendous parallel between soil and human health. Both are based upon a
balance of minerals and microbes and in each case this balance has been seriously
compromised. NTS is about regenerating this life force.

This course will be a prerequisite for growers to qualify for marketing under the proposed
"Nutrition Farming" banner. The NTS "Nutrition Farming" approach is being developed
as the industry standard for increased yield of nutrient-dense foods, which can attract
premium prices for longer shelf-life and forgotten flavours.

Appendix F—Mineral supplements for human health.

To be sure that your natural sources actually contain the required elements you need to grow your own food and ensure the soil they are grown in has all the elements. But each plant uses different elements so how many different fruits and vegetables do you need to eat to get all the minerals? It is certain that most soils are deficient in some minerals and plants cannot create those that are missing. Then here is a solution. You need to take a mineral supplement that is as natural as possible that has all the 92 elements in it. My internet research has found four possible overseas sources:

1. Dr. Joel Wallach at www.wallachonline.com has many supplements.
2. Quinton's Marine Plasma sold under the name Marine Matrix in the USA. Based on seawater with all the elements. See www.plasmaquinton.com & www.originalquinton.com & www.quinton-seawater.com .
3. T J Clark's colloidal spring water from Utah at www.tjclarkinc.com has over 70 minerals.
4. Megamin a European product ---www.megamin.net (150 capsules 750mg for $49 US or $80 Aus).

Sources in Australia.

1. TJ Clark's Colloidal Mineral Formula id available in some Health Food shops. 200ml glass bottle costs $19.95.
2. Quinton's Marine Plasma is available in Brisbane from Dr Greg Emerson for $90 for 24 ampoules which suggests it is injected into the bloodstream.
3. Dr.Wallach's supplements are apparently available in Australia too and I am trying to find where.

In addition to the above sea vegetables are another good source and as for soil health one of the best is kelp. Most health food shops stock kelp (or seaweed) capsules. Sea water and sea vegetables have very similar chemical composition to human body fluids. Millions of years of soil erosion has washed all the minerals into the sea. Sea vegetables are higher in vitamins and minerals than any other food containing all necessary trace elements for life. The main sea vegetables are:
Kelp, Kombu (a Japanese delicacy), Hijiki, Nori (used as a sushi wrapper), Arame, Sea Palm (from the Pacific coast of North America), Bladderwack, Wakame, Dulse and Irish Moss. (See reference 1).

Appendix G

"Growing Healthy" training courses run By Geoff & Bev Buckley

Geoff & Bev Buckley have over 25 years experience in education and training and have been farming successfully for the past ten years. We grow 60 different fruit and vegetables organically. We won the Queensland Small Business Champion Award for Fresh Food and were a finalist in the 2008 Australian Vegetable Grower of the Year Awards. We have been featured on TV and in magazines and the press regularly. In 2009 we were both awarded a Diploma in Horticulture.

We currently offer two training programs:

1. A twelve module monthly training series downloadable from the internet. See their two websites www.healthygardenfood.com and our blog http://healthyorganicnow.com . Each lesson has a companion video.

2. An evening and weekend practical training run on their farm where they demonstrate how to grow fruit and vegetables organically and you learn by trying it out yourself. This program runs over four evenings from 7.00pm to 9.30pm plus a one day workshop on a Sunday with a healthy delicious lunch provided.

We plan to build a residential training centre on our property and then will offer an introductory two day weekend workshop and a six day residential program.
The centre comprises ten cabins each sleeping two people plus a communal building with a commercial kitchen/dining room and training room. Meals will be provided to demonstrate the extraordinary range of tastes and nutritional benefits from fresh, local organic food. We believe you will come away healthier than when you arrive after experiencing the fresh air, clean water, fabulous food and the beauty of Mount Tamborine.
Until the centre is completed we run face-to-face training courses in an evening & weekend format; also we do consulting on your property if you cannot make our courses (contact us by email for more information at buckleyg@bigpond.net.au).

N.B. We offer a free monthly newsletter with recipes, nutritional information and discussion of relevant topical issues. You can obtain a copy sent by email by visiting the above websites or our new website www.growinghealthyorganicfood.com . We are also about to release a computerised "planting calendar" that tells you what crops to plant this month in **your** climate zone and **your** hemisphere. We are also writing (i) a new book on "Transition Farming" with 20 case studies of Australian Farmers who have converted from conventional to organic farming, and (ii) a new training course on how to double your income growing by teaching the business and marketing principles to maximize your return from growing plus a unique analysis of the profitability of over 130 crops.

Index

www.ingramcontent.com/pod-product-compliance
Lightning Source LLC
Chambersburg PA
CBHW041443210326
41599CB00004B/121